T0167163

Dear Jim:
Our History of ITIS

A Restructuring of Thought from Myth,
Fiction, and Institutionalized History,
Describing Homo Saps'
Development of
Information Technology (IT) and
Information Systems (IS),
And the Continuing Competition between
Individuals and Institutions
Over the Control of the ITIS Tools

John Barber

iUniverse, Inc.
Bloomington

Dear Jim: Our History of ITIS
A Restructuring of Thought from Myth, Fiction, and Institutionalized
History, describing Homo Saps' Development of Information Technology
(IT) and Information Systems (IS), and the Continuing Competition
between Individuals and Institutions over the Control of the ITIS Tools

iUniverse books may be ordered through booksellers or by contacting:

iUniverse
1663 Liberty Drive
Bloomington, IN 47403
www.iuniverse.com
1-800-Authors (1-800-288-4677)

Because of the dynamic nature of the Internet, any Web addresses or
links contained in this book may have changed since publication and
may no longer be valid. The views expressed in this work are solely those
of the author and do not necessarily reflect the views of the publisher,
and the publisher hereby disclaims any responsibility for them.

Any people depicted in stock imagery provided by Thinkstock are models,
and such images are being used for illustrative purposes only.

Certain stock imagery © Thinkstock.

ISBN: 978-1-4502-8608-4 (pbk)
ISBN: 978-1-4502-8609-1 (ebk)

Printed in the United States of America

iUniverse rev. date: 1/28/2011

CONTENTS

PREFACE

Dear Jim:

Here is Part I of Our History of ITIS. The set of Thought sequences that initialized our compilation of Thought structures in this WritingITIS product are those we sequenced in our discussion about information packaging techniques.

You will remember the occasion as our after-work meeting on one Friday in a hot August at least a Homo Sap-generation ago in the King George Tavern. What a wonderful suggestion you had that we copy the medical world by pronouncing this "eye-tis." We fuelled our discussion with a couple of cold lagers and superb Monte Cristo cigars.

Those were the days and they are long gone. Pity, eh!

We were both deeply involved in the business of packaging information and investigating developments in Information Technology (IT) and Information Systems (IS) as aids in the packaging process. In some convoluted way, as in any discussion of ITIS products and Homo Saps' instinctive adoption of user-friendly ITIS tools, we ended (because you had to go home to dinner) with a discussion of Prefaces and Forewords and their functions as separate components of WritingITIS products such as books.

Before we finally destroyed our cigars, we agreed—or at least, in your institutionalized way, left unresolved—that the cosmic god Thoth, the ultimate ITISan, prompted Homo Saps to develop the two information-packaging techniques to separate and disseminate different messages. He suggested Homo Saps design Prefaces to contain Thought structures with IS components disseminating messages the least relevant to the Thought structures, IS components, and messages of the primary WritingITIS products, and Forewords to contain Thought structures, IS components, and messages more closely related to those of the primary WritingITIS products. From these Thought structures grew Our History of ITIS.

FOREWORD

Dear Jim:

Our initial investigation into the functions of Prefaces and Forewords only exposed the mass of historical information describing Homo Saps' development of—some storytellers might say love affair with—user-friendly ITIS tools. The ITIS applications they designed for the tools—again under Thoth's godly guidance—focused on two primary objectives. One focused on civilizing the hominid animal part of Homo Saps. The other focused on enhancing the individual Thought-processing freedom of the Homo Sapiens part of Homo Saps.

Our sources of information kept leading us into processing the Thought structures that storytellers captured in ITIS products, mostly WritingITIS products. They labelled the IS components and messages myth and fiction, as well as historiography that followed institutionally approved protocols for sequencing historical Thought structures.

We identified five ITIS tools that Homo Saps have consistently developed, sometimes with great passion, over hundreds of Homo Sap generations. We labelled the tools, in order of their development, OralITIS, ImageITIS, CalendarITIS, WritingITIS, and AlphabetITIS.

We relabelled the "information" as "IS components and messages" embedded in "Thought sequences" completing the Thought structures that the ITIS products contained. We did this because much "information" about the ITIS tools, their development, and impacts on individual Homo Saps and their civilizations, appeared in Thought structures focused on other topics. Homo Saps have developed ITIS skills that allow them to hide such IS components and messages in their ITIS products to ensure the continued dissemination of the fundamental messages.

For example, in several ITIS products containing Thought structures primarily describing the actions of the Roman Homo Sap

Julius Caesar, storytellers introduce the Thought structure whose IS component describes his desire to open up the Roman libraries to any Roman citizen. But that is all they say. They omit any Thought structures explaining why he had such a desire. The hidden IS message indicated that the Roman institutions restricted access to the Thought structures of the ITIS products they stored in the libraries. But why did they do that? What prompted storytellers to keep repeating the Homo Sap Julius Caesar's desire? Did he succeed? Or, were they fearful of ordinary Roman Homo Saps acquiring ITIS and Thought-processing skills that would undermine the institutional control of Rome and unravel their self-serving institutional machinations?

Storytellers also claim he was a consummate ITISan and Homo Saps leader—first man of Rome, leader of the aristocratic, priestly, and military institutions of Rome. He was very much concerned with the welfare and rights of his people—the primary concern of all Homo Sap leaders. He had unsurpassed Thought processing and ITIS skills. He had a prodigious Homo Sap memory, a critical ITIS storage peripheral. He demonstrated great skills in producing ITIS products, both WritingITIS and OralITIS that other Homo Saps still study. He implemented a new version of the ITIS tool of CalendarITIS in the Roman Homo Sap civilization, effectively curtailing the power of the Pontificate and Senate, Ancient Roman priestly and aristocratic institutions, to control the sequencing or timing of Roman events to suit their institutional purposes. He understood from his experience as chief pontiff that CalendarITIS is primarily an institutional tool for regimenting the activities of individual Homo Saps.

I have structured this WritingITIS product—our book, that is—to capture and focus on Thought sequences whose IS components and messages describe Homo Saps' relationships with and development of the ITIS tools and with institutions that attempt to control the development and applications of the ITIS tools. I have selected the Thought sequences from ITIS products that both individual and institutionalized Homo Saps have produced. Whether fancy or fact, they are all the products of the Homo Saps Thought-processing imperative—an imperative as powerful as the hominid animal physical survival (HAPS) imperative.

From the WritingITIS products that describe myths, come the

Thought sequences describing Homo Saps' relationships with their gods, particularly the sexual ones, and the connections of the ITIS tools with the gods and godly Thought processes. From those of fiction come much of the Thought sequences describing the impact of user-friendly ITIS tools that freed individual Homo Saps from institutional domination—from those institutionally-approved histories come "facts."

Here are a couple of primary outcomes of the Thought connections between the mythical, fictional, and historical Thought structures:

Homo Saps are Earthly entities who have developed themselves, primarily through Thought processing in their individual Homo Sap memories, into two primary but interdependent species: One a hominid animal species and the other a Thought-processing species.

When the Homo Sap Aristotle died, the Ancient Greek philosophers had by then sequenced sets of Thought structures whose outcomes established that Homo Saps operated in the Earthly environment as two kinds of Thought-processing species: One an individual Thought species, and the other an institutional Thought species.

Both species operated continuously in every single Homo Sap, sometimes in harmony but often in conflict between the survival demands of each species. Both also had to align their survival demands with the overriding survival demands of the hominid animal species in which they operated and on which they depended.

The most important developments in the ITIS tools that Homo Saps adopted, instinctively and without question, have been those that made the ITIS tools more and more user-friendly to individual Homo Saps. This becomes most evident in Part II of Our History of ITIS: After Aristotle Died.

Homo Saps did have their standard killing sprees during the 115 or so generations following Aristotle's. Most of the killing sprees were those that institutions instigated with ITIS products containing IS components and messages that triggered Homo Saps' animal fear for HAP survival. These killing sprees changed the operating conditions of their Earthly environments and gave the prevailing institutions greater control over individual Homo Saps.

The most important changes in the Homo Saps' civilizations were those that individual Homo Saps began by learning ITIS skills and taking control of the ITIS tools and messages ITIS products disseminated.

1:
BEGINNINGS

Dear Jim:

Homo Saps are Earthly entities that operate as two species in the Earthly environment. Storytellers—a designation many Homo Saps aspire to—label one species whose operating imperative compels them to function as Thought processors, the "Homo Sapiens." They label the second species whose operating imperative focuses on Earthly physical survival "hominid animals." Homo Saps have to function in accordance with the directives of both imperatives, mostly symbiotically.

Homo Saps began operating in the Earth environment 40,000 Earth years ago, producing about 2,000 generations. They functioned primarily as hominid animals for the first 30,000 of those years, producing 1,500 generations of Thought-processors. Each generation developed greater Thought-processing power than the previous one, producing incrementally successful outcomes that improved their survival as a species. The successes allowed them to make themselves the dominant hominid animal species 10,000 Earth years ago. They made themselves the central Thought-processor units of their Earthly environments over the following 500 generations.

Their Thought-processing imperative, which prodded them continuously, directed them to make many changes to their Earthly environments and operating procedures. Most changes they initiated by developing their five information technology and information system (ITIS) tools, which were the inventions of a cosmic god they named Thoth. They developed the ITIS tools as application extensions of

1

their embedded hominid animal physical (HAP) ITs: vocal, visual, and aural.

They developed first the ITIS tools of OralITIS and ImageITIS and initialized the IT and IS specifications to focus their applications primarily on the survival of their hominid animal physical (HAP) structures. Next they developed CalendarITIS, which was primarily an institutional tool for directing and controlling the activities of individual Homo Saps. They developed WritingITIS with IT specifications that allowed them to produce Thought structures with IS components disseminating messages that promoted only institutionalized outcomes. Finally, they developed AlphabetITIS with simple IT specifications that allowed individual Homo Saps to produce WritingITIS products with Thought structures containing IS components that disseminated messages that challenged those of their institutions.

They developed applications of the ITIS tools to function on the platform of the five information-gathering technologies embedded in their hominid animal physical (HAP) structures. The five technologies, their vocal, visual, aural, taste, and touch technologies, interface with their HAP Operating Systems (HAPOS). Their function is to feed environmental and other information to HAPOS utilities that control HAP activities in the immediate Earthly environment.

Homo Saps designed many different versions of the ITIS tools. They fixed the designs with IT specifications and IS communication protocols. They updated the specifications and protocols, adding features and applications that helped them consolidate and establish their different civilizations.

The Homo Sap leader, an Ancient Egyptian, to whom storytellers gave the name Narmer, was the first to establish the precise specifications and protocols of Ancient Egyptian versions of the five ITIS tools. He designed and developed their applications to help him establish his civilization, the first stable and coherent Homo Sap civilization.

He implemented the first four of the ITIS tools and made himself the first Ancient Egyptian god-king. He specified a set of god-king operating procedures with more restrictive outcomes than those of his Homo Sap leadership operating procedures. The god-king operating procedures gave him absolute control over his Homo Sap group; he needed that to ensure the stable development of his nascent civilization. His Homo Sap leadership operating procedures focused only on the survival and welfare of his group and his tenure as group leader depended on the success of procedures he devised to keep his group safe and fed. He maintained the two sets of operating procedures in separate Thoughtbases that he stored in his individual Homo Sap memory.

Narmer's failure to implement the fifth ITIS tool, the Ancient Egyptian version of AlphabetITIS, was deliberate. His god-king analysis of AlphabetITIS identified fundamental stability problems for his fledgling civilization. His god-king Thoughtbase classified AlphabetITIS as dangerous to a god-king's absolute control of developments in a Homo Sap civilization. AlphabetITIS would allow individual Homo Saps to acquire independent WritingITIS skills, improve their independent Thought-processing capabilities, expand their individual Homo Sap memories, and produce WritingITIS products capturing Thought structures with IS components that disseminated messages challenging his absolute control of Thought processing, a critical attribute necessary for the stability of all nascent Homo Sap civilizations.

Narmer based his analysis on the Homo Sap leadership Thoughtbase that his predecessor transmitted to him in OralITIS products. The Thoughtbase contained the Thought structures his predecessors sequenced and stored in individual memory to describe the successes and failures of their actions. Virtually all his Earthly activities focused on storing the Thoughtbase in his individual Homo Sap memory, identifying the key-Thought for accessing the leadership procedures, and preparing himself to assume the leadership responsibilities of his group.

All his predecessors identified recurring problems they had to resolve. One was the challenges of individuals who had improved their Thought-processing capabilities. They had to eliminate some of those challenges by executing hominid animal killing procedures to eliminate the challengers. Another persistent problem, again the outcome of improved Thought-processing capabilities, was Homo Saps' inability to avoid sequencing Thought—the outcome of their Thought-processing imperative—into structures with IS components disseminating messages that explained their provenance, their beginnings.

THE FIRST BEGINNING—THOUGHT AND MEMORY

Storytellers who focus their Thought processing on Thought structures whose outcomes they label as myth, fiction, and science, generally agree on the events of the First Beginning. The outcomes of these events manifested themselves to Homo Saps as Cosmos, gods, and finally Homo Saps.

Storytellers describe the First Beginning as that when only THOUGHT was and still is. THOUGHT came before all that exists or existed. THOUGHT begins everything. All that ever was still exists as a unified whole, but only for THOUGHT and only in MEMORY, the store of all free-Thought.

The existence of THOUGHT, the precedence of THOUGHT and the criticality of MEMORY before anything else is clear in the IS components and messages that Homo Saps embedded in Thought structures many Homo Sap groups captured and persistently repeated in their OralITIS and WritingITIS products. One of the clearest networks of Thought structures disseminating these messages was one that Homo Saps, who lived in the cold northern regions of the Earthly reality, sequenced. The IS components identify extra-terrestrial entities that storytellers label "gods." The Thought structures describe the Homo Saps' habit of appealing to gods for direction on the safest Earthly procedures they should execute whenever they were fearful of the outcomes of life-threatening activities they were about to execute.

The critical Thought sequence containing the message describes how the chief god of the northern Homo Saps made his appearance: Out of the dense mists of dark northern forests a shadowy figure emerges. A hooded cloak hides his features and physical form. The shadowy figure is the northerners' chief god to whom storytellers gave the name "Odin." The sequence describes two ravens leading Odin out of the mists. Storytellers gave the ravens the names THOUGHT and MEMORY. The ravens, completely free, flew ahead of Odin, guiding him. THOUGHT always preceded MEMORY and pointed the way for the god.

The IS components disseminated other messages. The hooded, shadowy figure was a guise that Homo Saps' gods adopted 35 generations before the Homo Sap Aristotle died. Homo Saps had begun to lose their awe of the gods. The event coincided with the Ancient Greek Homo Saps' adoption of a simple 22-letter ITIS tool, AlphabetITIS, for capturing their Thought structures in WritingITIS products. For the following generations the aristocratic and priestly institutions, all of whose members claimed a special relationship with the gods and acted on godly direction, sequenced all godly Thought structures to contain IS components with messages describing gods as institutionalized, dark, extra-terrestrial entities with ominous, omnipotent, omniscient powers controlling Homo Saps' HAP survival in the Earth's environment.

The godly purpose was to generate fear, the outcome of a complex set of Thought structures that gods were most fond of prodding in the individual memory of each Homo Sap. The cosmic godly institution through their priestly and aristocratic Earthbound institutional surrogates forbade Homo Saps from looking at and seeing any godly features or of naming them, because using OralITIS products to utter the names of gods allowed individual Homo Saps to invoke the omnipotent and omniscient powers of the gods. The institutional purpose was to maintain the gods as omniscient, omnipotent entities; a status that Homo Saps' free-Thought-processing of Thought structures had been steadily eroding since the development of Ancient Greek AlphabetITIS.

The THOUGHT Imperative

THOUGHT had one overriding attribute and imperative: Processing the infinity of free-Thought in MEMORY was an unceasing activity with two objectives: One was to sequence free-Thought into structures whose outcomes described realities. The other was to re-sequence Thought structures to describe outcomes that modified existing realities or defined different realities.

The realities that THOUGHT defined in Thought structures, networks of Thought structures, and internetworks of networks, had two attributes: One was order; the other was uncertainty.

Uncertainty kept THOUGHT, and consequently Homo Saps, from precisely defining their realities. Free-Thought from MEMORY, always available to Homo Saps, always changed the Thought structures that Homo Saps sequenced to describe the realities. They stored the Thought structures in Thoughtbases that they established for random access and processing in their continuously operating Homo Sap memories.

They developed the ITIS tools they needed to capture and isolate Thought structures in ITIS products before new free-Thought corrupted original Thought sequences and modified outcomes. They developed the ITIS tools as Thought-processing apprentices to the ultimate

ITISan, the cosmic god Thoth, who invented the ITIS tools, much to the chagrin of the Earthbound god-kings.

The cosmic god Thoth ignored the god-kings' complaints.

Homo Saps' beginning was an outcome of THOUGHT sequencing free-Thought into a two-dimensional structure with two outcomes. One dimension specified operating procedures focused on fixing the order of a reality and the other dimension specified operating procedures focused on controlling the uncertainty in the reality. In one reality, the order and control Thought structure initiated a series of Thought events whose outcomes were both real in the reality and, as Thought in MEMORY, independent of the reality. That was a paradox: The reality was both a real, sensible system and an unreal, un-sensible system.

Of all living things, only Homo Saps tasked themselves to access MEMORY, to retrieve and sequence Thought into structures whose outcomes could explain their provenance and the provenance of all living things and realities. They developed both bottom-up and top-down Thought analysis techniques to fill in the Thought sequences from the First Beginning.

The cosmic god Thoth could influence Homo Saps in only one way. He could prompt their Thought-processing capabilities—capabilities most of the cosmic gods lacked. He did have access to the godly Thoughtbase for procedures that could change physical events in the Homo Saps' Earthly environment, but his elder brother, the cosmic god Ra, forbade him from using his godly powers. The cosmic god Ra was the Sun god, the supreme godly supervisor of the Solar System with absolute control over all physical events in the Solar System. Homo Saps owe their provenance to the outcomes of a couple of godly disputes between Ra and Thoth, outcomes that left Thoth partly responsible for the welfare of Homo Saps.

Thoth fulfilled his responsibility by prompting Homo Saps to develop the five ITIS tools to aid them in their Thought processing. He focused his efforts on the development of the last two ITIS tools, WritingITIS and AlphabetITIS to help them clarify the outcomes of their Thought processing and give each generation of their Earthly offspring Thoughtbases to continue expanding.

Homo Sap leaders needed only the first four ITIS tools to form, establish, control, and stabilize Homo Sap civilizations, the first of which was the Ancient Egyptian civilization. Without the four ITIS tools, Homo Saps operated in the Earthly environment as simple hominid animals.

Homo Saps sorted the outcomes of their Thought processing and selectively connected Thought structures to reach outcomes that defined their Beginning. The bottom-up analyses identified immediate procedures to control immediate events that impacted their physical survival. They began top-down analyses with a starting Thought, signalling a new Beginning, to connect their Thought into unified structures whose outcome provided them with a plausible provenance.

Their Thought-processing imperative compelled them to retrieve free-Thought from MEMORY and store the Thought in the free space of their expandable individual memories. They immediately melded the new Thought into the sequences of existing Thought structures to improve the plausibility of outcomes and embedded messages. The Homo Saps' Thought-processing imperative compelled them to keep sequencing Thought structures that explained their reality.

Homo Saps have for hundreds of Homo Sap generations sequenced and processed Thought structures in the infinite free space of individual Homo Sap memory to improve both the IT and IS components of their ITIS tools, with special emphasis on the ITIS tool of WritingITIS. The outcomes of these ITIS advances have always increased the freedoms of individual Homo Saps and diminished the onerous powers of their institutions to control their Thought processing.

Two thousand generations of original Homo Sap Thought processing ended with that of the Homo Sap Aristotle. He and his Ancient Greek predecessors captured and re-sequenced Thought structures containing IS components disseminating powerful messages. They were fortunate enough to have available to them a new version of WritingITIS. The Homo Saps' Thought processing after the Homo Sap Aristotle died focused on the development of the IS components and messages the Ancient Greek Homo Saps captured in their WritingITIS products.

The Ancient Greek Homo Saps had been developing—without institutional interference—their version of WritingITIS using the simple 22-letter version of AlphabetITIS and demonstrated a usable version 25 generations before Aristotle's. The features of their WritingITIS allowed them to distil, simplify, and clarify the outcomes of the previous 2,000 generations of Homo Saps' Thought processing. The Thought structures they produced contained IS components disseminating powerful messages.

One of the most powerful described Homo Saps as three distinct species operating in one Earthly physical structure.

One species is the hominid animal species. This species has limited Thought-processing capabilities; all focused on the survival on their HAP structures.

The other two are Homo Sap Thought species, who operate independently of their HAP hosts. They are completely dependent for survival on the efficiency with which the hominid animal species access their physical survival Thoughtbase and retrieve and execute physical survival procedures. They are also dependant on the five information-gathering technologies embedded in the HAP structure.

The two Homo Sap Thought species are the individual Thought species, and the institutional Thought species. Each of the Thought species seek similar Thought-processing outcomes but contend with each other over which species should be the dominant one and over the freedom of individual Homo Saps to sequence and process free-Thought. Both identify the user-friendliness of the five ITIS tools as critical to the resolution of their conflict.

2:

HOMO SAPS DEVELOP THE FIRST TWO ITIS TOOLS: OralITIS AND ImageITIS

Dear Jim:

The Thought processing of the first 1500 Homo Sap generations initiated a series of ITIS events with incremental outcomes, each enhancing previous ones. The outcomes—among them, their development of the first two ITIS tools of OralITIS and ImageITIS—allowed Homo Saps to establish themselves as the dominant hominid animal species operating in the Earthly environment.

The domination process was more the outcome of a survival necessity than a definite plan. The simple hominid animals ejected Homo Saps from the hominid animal groups in which they sheltered, denying them the protection of the large groups. The ejection was the outcome of Homo Saps' inability to control or suppress the demands of their Homo Sap Thought-processing imperative, a unique operating feature that distinguished them from simple hominids and other animal species. A traumatic event for Homo Saps, the ejection forced them to form their own groups, seeking the protection of large numbers from predators.

Traumatic Earthly events that Homo Saps had to confront in their development as Earthly entities forced them to develop their Thought-processing capabilities and sequence procedures that enhanced the paramount survival needs of their HAP structures. Thought processing expanded the storage capacity and independent Thought-processing facility of their individual Homo Sap memories. The outcomes led to their initialization of the specifications of the first two ITIS tools, OralITIS and ImageITIS. They designed the first applications of these ITIS tools as survival applications. Their successful outcomes began to give them the competitive advantage over all other animal species.

The cause of their ejection was their random processing of Thought structures other than those in the HAP survival Thoughtbase. Random Thought processing distracted them from their HAP survival responsibilities and exposed the groups to attacks from predator animal species.

The majority members of the original hominid animal groups, all simple animals with a Thought-processing facility capable of identifying only binary physical survival outcomes, began to classify their Homo Sap members as survival threats. They identified any individual hominid animal that exhibited characteristics other than those of the majority simple hominid animals as Homo Saps. The simple hominids ostracized them, abandoning them to fend for themselves, often to die.

Homo Saps who survived as sustainable groups took advantage of their expandable individual memories to resolve the often conflicting demands of their HAP survival imperative and their Homo Sap Thought-processing imperative. They began to design new Thoughtbases and store them in the free space outside the closed HAP survival Thoughtbase. The HAP survival Thoughtbase stored survival procedures with only binary outcomes: Live or die; kill or be killed.

They retrieved HAP survival procedures and stored them in the new free-Thoughtbases. They reprocessed the retrieved Thought structures, and the new open Thoughtbases allowed them to add sequences that offered them a wider range of survival options than the two-valued ones open to hominid and other animals. The new Thoughtbases allowed them to begin sequencing Thought structures with IS components, distinguishing them from pure hominids.

They began to design IS components for the new Thought structures to embed and disseminate messages identifying Homo Saps features distinguishing them from other animals. One that had a powerful influence on their Earthly survival identified Homo Saps as two species operating symbiotically in one physical structure: an animal species and a Thought-processing species. Homo Saps had to function within the physical limitations that the Earthly Operating System (EOS) imposed on the operating systems and technologies of their HAP structures.

Individual hominid animals stabilized as independent Earthly physical entities with features that two sets of uncertainties helped fix. One set Homo Saps identified as the range of possible outcomes of the cosmic operating system (COS) and EOS operating utilities. The primary function of these utilities is to maintain equilibrium and balance in the Earthly environment. The COS and EOS execute the utilities automatically whenever any event changes the immediate equilibrium, forcing both cosmic and Earthly environments into states of constant change. The outcomes modify hominid as well as all other animal physical structures as they grow into viable adult animals. The modifications give individuals the capabilities to operate and sustain themselves as animals under the conditions of their immediate Earthly environments.

The second set identifies the uncertainties of the outcomes of animal sexual procedures: Only one of them is the reproduction procedure with a binary outcome. The complex interaction of the outcomes of COS and EOS utilities with the sexual reproduction procedures during their execution arbitrarily produce either male or female individuals. The offspring of those who are of the Homo Sap species also stabilize with expandable memories and varying Thought-processing capabilities. All individual animals stabilize with at least one feature that makes the individual unique.

All species incorporate individual animal operating systems (AOS)—each a very limited subset of the COS and EOS. The AOS controls the mechanical and other functions of their Earthly physical structures. AOS driving utilities automatically activate technologies of peripheral devices that individuals need to accomplish survival outcomes. Some utilities operate continuously to allow individuals to adapt to and function within the range of physical conditions that the uncertainties of EOS outcomes establish in the Earthly environment in which they operate. Some utilities operate to activate and control embedded electro-mechanical and chemical technologies that maintain the operating viability of their animal physical structures.

EMBEDDED IT AND IS APPLICATIONS

All hominid animals stabilize with four information-gathering technologies and one information-transmitting technology; all five installations are original design features of their Earthly physical structures. They gather information with their visual, aural, taste, and touch technologies. They transmit information with their vocal technology. The primary functions of these embedded technologies are to ensure the physical survival of each individual animal.

AOS utilities activate the visual and aural technologies to poll their immediate physical environment for data describing events and conditions that threaten physical survival. AOS sub-routines convert the polling outcomes into ImageITIS and OralITIS products. The IS components contain messages that define the threat level of the environment as hostile, neutral, or benign. ImageITIS or OralITIS products disseminating messages of a hostile environment immediately activates appropriate survival procedures in the animal species Thoughtbase. Each animal species executes survival procedures compatible with their physical operating capabilities.

The input ports for the giga-quantities of environmental data defining the ImageITIS and OralITIS products are the hominid animal eyes, ears, tongue, and skin. The polling applications contain sub-routines that convert the environmental data into information that other subroutines capture in the IS components of the Thought structures defining the two sets of ITIS products. The IS components disseminate fixed messages that alert individual hominid animals to the threat level and other survival conditions in their immediate environment. The survival Thoughtbase stores all messages and links them to their associated survival procedures. The threat levels compel each individual to retrieve and activate appropriate survival procedures.

Homo Saps Design Their First OralITIS Products and Communications Protocols

The outputs of the vocal technology are sounds that other animals can capture with their aural technologies. Each animal species can activate a vocal technology that produces different ranges of sounds. The applications condition the technology to allow each animal to produce sounds—crude OralITIS products—that contain IS components disseminating messages to which other individuals of the same species can respond to with specific reactions. Most of the messages confirm or inform other members of the animal species of specific survival conditions. Alerts to dangerous survival conditions force individual animals to activate an application that produces loud, raucous sounds that force other individuals to cease their activities and activate their defensive or offensive physical survival peripherals, readying them to execute survival procedures.

Homo Saps sequenced the sounds into structures that defined their first OralITIS products. They identified technical qualities—volume and tone—of the sounds. They associated the technical qualities of the sound with the threat level and established the first communications protocols—very loud, harsh screams, yells, or roars identified and broadcast very high threat levels.

HOMO SAPS ASSIGN SURVIVAL RESPONSIBILITIES

Most animals process Thought structures exclusively in their survival Thoughtbases: Their animal physical survival imperative is in absolute control, always instantly ready to react to physical threats. Homo Saps had a deadly option: They could process Thought structures exclusively in their free memory, free themselves from all Earthly constraints, make themselves oblivious of all dangers, and ignore the demands of their HAP survival imperative. The outcomes were often fatal. Their Homo Sap Thought-processing imperative could unobtrusively redirect their Thought processing from their HAP survival Thoughtbase to the free space of individual memories and take complete control, which was euphoric, an addiction which storytellers label "day dreaming."

Thousands of generations of Homo Saps passed—mostly as the survival food resources of predators—before some individuals, their frequent Thought processing having expanded their free memory capacity for processing new Thought, began to analyze their problem. They began to sequence and process Thought structures whose IS components, messages, and outcomes focused on resolving their dilemma with their two operating imperatives: HAP survival, and their Homo Sap Thought-processing imperative.

One set of new Thought structures contained procedures directing them to assign individual Homo Sap members of their groups to the task of exclusively operating mostly in the free space of their individual memories. They made an agreement with these individuals. They freed them from most HAP survival activities and guaranteed them their HAP survival requirements. Their new survival functions were to act as surrogate free-Thought processors for the rest of their group and allow the other members to focus on survival activities.

The agreement required these designated individuals to disseminate the outcomes of their free-Thought processing to the rest of the group during the dark periods that stopped them from executing most survival activities. The groups used the dark periods to re-energize their HAP structures with survival food resources. HAPOS maintenance utilities repaired damage to their HAP structures from physical survival activities during the light periods. They transmitted their Thought structures and received Thought structures from other individuals while the groups lay inactive around their fires.

Storytellers label these periodic events "Rest and Re-creation." They labelled the designated individuals "storytellers."

THE ITIS OUTCOMES OF THE FIRST STORYTELLERS

The storytellers began to design new applications for their HAP vocal technology. The grunts, screams, roars, and moans of their primitive survival OralITIS products were inadequate to capture and describe the IS components and messages of the Thought structures they sequenced. They began to develop new sounds and protocols for sequencing the sounds into OralITIS products that transmitted the IS components and messages. They began to develop ImageITIS products with which to physically demonstrate or illustrate the messages.

Their predominantly hostile environments kept interrupting their Thought processing. The small Homo Sap groups lacked the numbers of individuals capable of adequately protecting them from all immediate Earthly dangers. Paradoxically, the Homo Saps Thought-processing imperative, once a danger to their HAP survival procedures with random Thought processing, now directed the storytellers to focus their Thought processing on practical HAP survival outcomes.

The Thought-processing imperative directed the storytellers to retrieve Thought structures from the HAP survival Thoughtbase and store them in the free expandable space of their individual memories, freeing the survival Thought structures from the Thought-processing restrictions of the HAP survival Thoughtbases. The storytellers sequenced free-Thought and melded the sequences with the retrieved HAP survival Thought structures. The absence of Thought-processing inhibitions in free individual memories allowed them to re-sequence and reprocess the melded Thought structures and identify new survival outcomes. One critical outcome was the continuous expansion of storytellers' free memory space with continuous Thought processing in the free space.

They began to develop new Thought-processing techniques. Storytellers label them "system analysis applications." They designed the applications to identify the changes their free-Thought-processing produced in the HAP survival Thought structures and their outcomes. They identified some changes in the procedures as impossible for the operating utilities driving their HAP technologies and peripherals to execute. They began to identify the limits of the technologies and peripherals: the strength and flexibility of arms and legs, the capacity of their lungs, and the distances over which their vocal, visual, and aural technologies were effective. They identified changes that required individual Homo Saps to sequence Thought into structures whose procedures directed their HAP technologies to execute modified optional procedures instead of reflexively executing the simple, two-valued HAP survival procedures.

Some changes enhanced their HAP survival outcomes. They captured the enhanced procedures and stored them in new, open, expandable Thoughtbases in free memory. The storytellers transmitted the new Thoughtbases to their group members during the dark periods. They retransmitted them until all individuals had captured and stored the Thoughtbases in individual Homo Sap memories. They directed their groups to keep reprocessing the Thought structures until each member made them automatically accessible, but optional, user-friendly survival procedures.

The storytellers began to identify the visual and aural environmental prompts that forced their HAP operating utilities to automatically and reflexively retrieve and execute their HAP survival procedures. They identified the primary prompts as specific visual and aural data that the limited Thought processing of their co-existing hominid animal species sorted from the giga-quantities of environmental data that they continuously and automatically gathered with their eyes and ears. They updated the analysis applications to capture and sort other environmental data to describe in more and more detail their immediate operating environment. Sub-routines in the applications processed the data in the free space of individual Homo Sap memories to produce Thought structures whose IS components began to explain the provenance and sequences of Earthly physical events whose outcomes prompted the reflexive physical survival reactions of simple animals.

Freely processing the new Thought structures during their R&R periods, individual members of their groups satisfied the demands of their individual Thought-processing imperative. The storytellers' Thought structures relieved them of the stresses of forcibly suppressing their own Thought-processing imperative to focus during the light periods exclusively on their HAP survival activities. They began to anticipate the relief, the anticipation prompting them to execute their physical survival activities with greater effectiveness.

The processing of the storytellers' Thought structures helped individuals Homo Saps expand their own memories. Some began to add the sequences to their individual survival procedures, improving outcomes. They could choose to either confront or avoid threats. Storytellers label these binary HAPOS outcomes "fight or flight." The closed animal survival Thoughtbase limited their choices to these two options. The new expanded Thought structures offered individuals many more options.

Homo Saps Begin to Focus on Group Survival and Initialize Institutions

Each generation of storytellers transmitted their expanded Thoughtbases to the next generation in OralITIS and ImageITIS products they designed to give the new generation sets of foundation Thoughtbases to reprocess and expand. They began to focus on sequencing Thought structures whose IS components described coordinated group survival activities. The change of focus from individual to group survival procedures forced them to develop new OralITIS products with new protocols for communicating with groups of Homo Saps. They connected Thought structures into narratives, describing past events during which the actions of the group in responding to threats impacted the survival outcomes. Storytellers labelled them "histories."

These OralITIS products connected Thought structures whose IS components described sequences of events and the groups' responses to each event. They described how the outcome of each event impacted the next event and how the group could have responded to ameliorate the negative outcomes of each event or the series of events.

They identified environmental conditions that always preceded the onset of events. They designed new OralITIS products with IS components that alerted the groups to prepare themselves to respond to a coming event. This was a momentous Thought processing event. Homo Saps had initialized a Thought-processing facility for predicting imminent events and developing the procedures for avoiding or taking advantage of them. All other animal survival Thoughtbases contained Thought structures that an animal executed only when the events were actually happening.

For disastrous events, storytellers developed procedures that directed their groups to prepare defences or take shelter before the events began. They identified the visual data that described the offensive or defensive procedures of attacking species. They developed Thought structures whose IS components described the sequences of physical actions that each attacking species executed in response to the defensive actions of the Homo Sap groups. They described precursor physical stances of the attacking species which signalled the onset of actions. Individuals and groups could then use, reconfigure their HAP structures, or sequence new procedures to defend against the actions before the attacking species began their actions.

The storytellers' new OralITIS products initiated fundamental developments in the social organization, growth, and safety of the Homo Sap groups. Processing the Thought structures containing IS components describing future events prompted more and more Homo Saps to practice disciplined Thought processing. They began to identify outcomes they wanted to achieve and practiced sequencing Thought structures with new procedures that allowed them to implement the outcomes.

The uncertain outcomes of the operating utilities that controlled the Earthly physical environment, combined with the uncertain outcomes of animal sexual procedures, produced individual offspring with a wide range of physical capabilities. Some offspring exhibited male characteristics, others female, the numbers of each arbitrary and unpredictable. The features of their HAPOS and HAP structures defined their differences. The Thought-processing capabilities in both sexes were identical. They all sequenced and processed Thought structures in individual Homo Sap memory.

Among the hominid animal species, some individual offspring grew into large HAP structures and exhibited physical strengths that allowed them to protect their groups against competing species and predators with greater success than those individuals in smaller, weaker HAP structures. The weaker individuals fell prey to predators more often than did the stronger ones, allowing the bigger, stronger ones to achieve greater survival rates. The groups with the bigger, stronger individuals in the majority tended to produce bigger, stronger offspring. The average physical size of the HAP structures of Homo Saps increased in the surviving Homo Sap groups with each new generation.

The Homo Sap females functioned in HAP structures that provided them with a unique survival capability. Only Homo Sap females could host, nurture, and deliver new individuals to sustain and increase their group numbers, critical requirements for the continuing survival of the species. Females of all HAP sizes and strengths carried and delivered offspring successfully. This exclusive reproduction capability also made the Homo Sap females more vulnerable to environmental dangers.

The HAP reproduction process allowed only females to host the new individuals in their physical structures. They had to carry them until their HAPOS utilities grew the offspring into HAP structures capable of surviving outside the female hosts. The females then had to execute procedures for sustaining, nurturing, and protecting the offspring until their HAPOS utilities had grown the offspring to their optimum HAP sizes with HAPOS capable of allowing the offspring to function as independent members of the groups. The HAP survival imperative compelled Homo Sap females to focus exclusively on nurturing and ensuring the survival of their offspring for longer periods than many other animal species.

This exclusive survival capability also influenced their sequencing and processing of survival Thought structures. While their offspring remained incapable of surviving as independent individuals, the female survival focus differed radically from that of the males.

Their functions as offspring hosts made the Homo Sap females less capable of helping in the protection of the groups to which they belonged. Having to carry the offspring in their HAP structures impaired their HAP technologies in escaping from, avoiding, or defending themselves against survival dangers. The extra weight they carried made them more vulnerable to physical damage.

Homo Sap females fought for physical survival with the same dedication and ferocity as did the males, but their specialist capabilities for producing new individuals had greater group survival value than their defensive physical capabilities. Homo Saps sequenced Thought structures containing IS components disseminating messages that connected the numbers of females in the Homo Sap groups with the sustainability and survival of the groups. The greater the numbers of females, the more offspring they produced to survive and grow into individuals capable of sustaining the numbers the groups needed to continue surviving.

These critical survival outcomes prompted Homo Saps to sequence Thought structures describing females as valuable group assets with specialist survival capabilities. The messages in the IS components emphasized the focus on group survival and the need to assign specialist functions to other members of the groups.

They developed procedures directing Homo Sap males to begin providing special protection for group females and their offspring. Homo Sap males positioned themselves as the first line of defence between the dangers and the female members of the groups in any confrontation with predatory animal species or other competing Homo Sap and hominid animal groups. The new procedures also directed the females to defend their offspring as their group survival responsibility.

As the first line of defence against survival threats, males exposed themselves to physical damage from attackers with greater frequency than did females. The losses of male members exceeded those of female members. The smaller, weaker males also succumbed to attacks more frequently than the bigger, stronger individuals.

The new female defence priority allowed them to subdue their HAP survival imperative and sustain extreme physical damage in the defence of their offspring. The outcome of this fierce dedication also had an impact on group survival. The loss of mothers left surviving offspring without immediate protection. Most Earthly animal species left motherless offspring unprotected and easy prey for predators. The survival value of offspring prompted the Homo Sap groups to develop new social procedures directing surviving females to care for and protect motherless offspring.

The new procedures were the first to give Homo Saps a competitive advantage over all other animal species. The outcomes impacted future social and operating procedures of the Homo Sap groups as well as their immediate survival. The HAP survival imperative still directed most of the reactions of both Homo Sap males and females, but many of the new Thought structures explicitly recognized group survival as a guarantee of individual survival.

The successes of the storytellers free-Thought processing stimulated free-Thought processing in the growing numbers of individual Homo Saps. They developed a Thought-processing discipline. They stopped inhibiting their Homo Sap Thought-processing imperative. They began to practice a Thought-processing skill that allowed them to focus the power of the imperative on enhancing their individual HAP survival procedures. The skill allowed them to store non-survival Thought sequences in a dedicated Thoughtbase. They disciplined themselves to access the Thoughtbase during their R&R.

HOMO SAPS BEGIN TO IDENTIFY THEIR UNIQUENESS

Homo Saps began to identify their individuality as Earthly operating entities when their Thought processing expanded as they released their Homo Saps Thought-processing imperative to process their storytellers' Thought structures during R&R. Their individual HAP technologies, different in detail from those of other individual Homo Saps, allowed some individuals to execute some survival procedures faster and with better outcomes than could other individuals. They began to identify specific survival procedures as more user-friendly to individual Homo Saps with specific sets of HAP capabilities. Individual Homo Saps began to reprocess survival procedures most user-friendly to individual HAP skills and identify other individual Homo Saps with similar HAP capabilities.

Though all Homo Saps functioned in HAP structures with HAPOS utilities driving the same sets of technologies, the uncertain impacts of combinations of COS and EOS utilities on the features of HAP structures as they grew to their optimum sizes, modified peripheral operating devices that allowed individuals to implement the technologies with different efficiencies and effectiveness. Some individuals grew with arms, legs, muscles, lungs, and other HAP features more capable of sustaining specific survival activities than other individual Homo Saps.

Homo Sap females sequenced the Thought structures connecting the survival of their offspring to the presence of male Homo Saps with big, strong HAP structures and operating peripherals. They had yet to make the connection between the execution of their sexual procedures with males and the development of new individuals in their HAP structures. Most of the offspring the females produced when the big, strong males conjugated with them also grew into big, strong individuals. The numbers of surviving individuals increased.

Prompted by the successes of their specialist storytellers, Homo Saps defined further procedures for assigning more specialist survival functions besides those of the production of new individuals and of defence. The messages in the Thought structures identified the greater productivity of cadres of Homo Saps who dedicated themselves to the group's survival activity. These cadres assumed responsibilities for survival activities, such as hunting for food resources and shelter construction. They implemented these specialist procedures and their numbers grew as more and more individuals survived without damage to their HAP structures from predator or other environmental dangers or the shortage of a survival resource.

Organizing their groups and building safe havens allowed Homo Sap groups to reduce the need for all group members to be constantly alert to extreme physical dangers. The HAP survival imperative took absolute control of the Thought processing only of those to whom the groups assigned the responsibility for providing the first line of defence against predators or other Homo Sap groups attacking them. This separation of the groups into dedicated cadres initialized the Thought structure with IS components describing new Homo Sap social entities. Storytellers labelled them "institutions." They labelled the formation process "institutionalization."

Each specialist group assured the others of specific survival needs. They institutionalized themselves. They developed procedures and protocols that controlled the membership to those Homo Saps who dedicated themselves to the purposes of the institutions.

SAFE HAVENS

The protected social environment—safe havens—gave Homo Sap individuals, those free of immediate defensive responsibilities or other institutionalized activities, more opportunities for their individual Thought-processing imperative to operate free of the inhibiting reactions of their HAP survival imperative. Homo Saps executed most survival routines reflexively, requiring minimal processing of the procedures. Their HAPOS utilities automatically executed them. Once individual Homo Saps captured new survival procedures and stored them in their individual Homo Sap memories, the drivers of their HAPOS utilities, operating in the background, always ready to run, activated the HAP technologies and completed the routines. This left the Homo Sap Thought-processing imperative free to sequence and process other Thought structures. Homo Saps labelled these routines "instinct."

The Thought-processing imperative took complete control of some individuals while they were executing instinctive survival routines. They were unable to stop themselves from sequencing and processing Thought structures with unreal outcomes in their free, uninhibited individual memory space. Many described physical environments that differed radically from their immediate Earthly reality. Some described fantastic environments. Those that captured Homo Saps' focus described environments in which they operated in complete safety from all physical harm or that were full of dangers they were unable to avoid.

Some individuals sequenced and processed Thought structures describing features like those of their real Earthly environments but with improvements that made the existing features more user-friendly or safer. These individuals stored the improved features in Thoughtbases and began to sequence Thought structures with practical procedures for recreating the improved features in their real environments. Storytellers labelled these individuals "artists," "artisans," and "inventors."

Other individuals sequenced Thought structures with IS components disseminating messages that automatically activated HAPOS utilities. Processing the Thought structures and messages in the privacy of their individual Homo Sap memories, they relieved themselves of the constant stresses of survival activities. The utilities forced their HAP structures to relax, relieving physical tensions. Some relief events allowed them to resume their survival activities with renewed physical energy. Some obliterated all Thought structures describing the dangers of their Earthly reality, the intensity of the relief compelling them to ignore their HAP survival imperative directing them to stop processing the Thought structures describing fantastic, unreal environments and return to processing Thought structures of their real Earthly environment.

These outcomes had a great impact on Homo Saps' Thought-processing developments. They stored the Thought structures describing both the fantastic environments and the stress relief they obtained from accessing and processing the Thought structures in independent Thoughtbases in Homo Sap memories.

They began to access these Thoughtbases and process the Thought structures with increasing frequency, seeking the physical stress relief again and again, neglecting their HAP survival responsibilities, oblivious to the events of their real Earthly environment. Group survival began to deteriorate with an impact similar to that that compelled the simple hominid animal groups to banish Homo Saps from the safety of the groups.

Homo Saps resolved the problem by directing their storytellers to form a cadre that focused on explaining these new and fantastic memory environments. The new storytellers designed and expanded OralITIS and ImageITIS protocols and procedures that captured the unreal Thought structures.

The Homo Sap audiences began to question why they were able to process the Thought structures, where the Thought structures came from, and why some of the unreal ones made them fearful, relieved their physical stresses, or made them ignore, at least temporarily, all physical dangers. The new storytellers sequenced Thought structures with IS components disseminating messages that crafted answers. They claimed that the source of their Thought structures were unreal extra-terrestrial entities they labelled as "gods." They labelled themselves "priests," privileged Earthly representatives of the gods who could communicate with the gods for explanations of any event. They claimed that the gods were the creators of all things, including Homo Saps, and with unEarthly magic controlled the fates of all things.

3:

A Godly Pre-History for Homo Saps: Big Bang, Singularity, Cosmos, Atum, Shu, Tefnut, Geb, Nut, Ra, and Thoth—the Ultimate ITISan

Dear Jim:

The Homo Sap Thought-processing imperative spawned storytellers of all kinds. They designed Thoughtbases in which to isolate and store in their individual Homo Sap memories sets of Thought structures with interrelated outcomes. They labelled the outcomes myth, fiction, science, and godly. They designed the Thought structures to contain IS components which disseminated many messages, many embedded in each other. The messages speculated on the provenance of Homo Saps and their Earthly and cosmic environments.

One set described Homo Saps as operating in the Earthly environment as a distinct species of hominid animal for 2,000 generations, spanning 40,000 Earth years. In this set they embedded another set describing Homo Saps as making themselves the dominant hominid animal species at the beginning of the last 500 of those 2,000 Homo Sap generations.

They described Homo Saps, as well as other animal species, as the arbitrary outcomes of the interactions of physical and chemical laws that determined the operating conditions of their Earthly environment. They specified these laws as the outcomes of COS and EOS utilities, driving the constant development and changes in cosmic and Earthly physical environments.

The first messages attributed all outcomes, real and unreal, to the actions of gods. They defined the gods as extra-Earthly entities—sometimes fickle, sometimes benign, sometimes wrathful—which could operate in the cosmic and Earthly environments without restrictions. The gods controlled the supervisory systems that monitored the conditions of all operating environments.

The reality and unreality of these messages existed only as the outcomes of Thought structures that Homo Saps sequenced and continuously processed in the Thoughtbases they stored in individual Homo Sap memories. The Thought-processing imperative allowed individual Homo Saps to freely re-sequence Thought structures to modify or eliminate messages and their outcomes in the privacy of their individual memories.

Storytellers in each Homo Sap generation captured with their available ITIS tools the Thought structures that previous generations sequenced and disseminated. Some storytellers reprocessed the Thought structures, instead of repeating them unchanged. They discarded some sequences, added more, expanded some and isolated others to produce Thought structures with modified or new outcomes that they disseminated to the other Homo Saps.

Storytellers identified many Beginnings for Homo Saps, all except one of which Homo Saps initiated. The one exception was the First Beginning. Only THOUGHT existed in the First Beginning, had and has one overriding attribute: to sequence Thought, the elements of THOUGHT, into structures defining realities. THOUGHT created MEMORY to store the outcomes of the Thought structures.

THOUGHT had and has three more immutable attributes: Two forced THOUGHT to order and control Thought sequencing and outcomes in MEMORY. The third attribute forced THOUGHT to sequence free-Thought in MEMORY and re-sequence stable Thought structures to determine new outcomes or modify existing outcomes. This Thought processing was continuous, without end.

MEMORY is an open system. Free-Thought spontaneously joins or merges with Thought structures, forming and reforming the Thought structures to reach new outcomes or modify old outcomes, without direction from THOUGHT—an attribute that individual Homo Sap memories acquired and which caused great confusion and dilemmas in the outcomes of Homo Sap Thought processing. Homo Saps developed the ITIS tools to capture and isolate Thought structures and help themselves resolve the confusion and dilemmas when the Thought structures disrupted their processing of other Thought structures.

BIG BANG, SINGULARITY, CHAOS

THOUGHT merged two Thought structures in MEMORY to initiate the First Beginning. Storytellers labelled the outcome of one structure "Big Bang," and the outcome of the other "Singularity."

The Big Bang Thought structure described Big Bang as a very hot, very fluid, intense ball of energy, tightly bound and stable.

The Singularity Thought structure described Singularity as a very dense, very heavy, infinitely small ball of bound matter.

One set of Thought sequences in each of the two structures defined the components of Big Bang and Singularity. Another set defined the procedures that THOUGHT sequenced to order and control the outcomes of Big Bang interacting with Singularity in MEMORY.

The merging of Big Bang and Singularity disordered the sequences of their Thought structures and disrupted the controls stabilizing the structures. Without controls, the Thought elements of Big Bang and Singularity functioned as free-Thought in MEMORY. They formed Thought sequences that changed constantly, continuously forming new but ephemeral Thought structures that defined a Thoughtbase with constantly changing outcomes. Storytellers labelled the Thoughtbase "Chaos."

The Chaos Thoughtbase defined a reality. The Chaos OS spontaneously broke the unstable Thought sequences apart and just as spontaneously reformed them into new sequences. Storytellers defined Chaos as an amorphous physical system with only disorder as an outcome. The physical components, free of the stabilizing controls of Big Bang and Singularity, joined together, burst apart, rejoined, burst apart again, and again. Each joining initialized Thought structures whose outcomes defined distinct but ephemeral subsets of realities, all possible in and out of Chaos in MEMORY.

Storytellers—expanding the Thought structures of 2,000 generations of their predecessors who focused their Thought processing on developing Thoughtbases whose outcomes they label "science"—

identified a still growing number of the physical components of Chaos. They gave them labels such as strings, up, down, charmed, strange, bottom, red, green, blue, cyan, magenta, yellow, white, quarks, antiquarks, gluons, streaking photons, negatives and positives, fractional negatives and positives, mesons, barons, leptons, and hadrons. All were outcomes of Thought networks that Homo Saps constructed in their search for beginnings. Spontaneous interactions between components filled the Chaos environment with forming, disintegrating, and reforming sequences whose outcomes produced dazzling displays of particles and light, changing constantly one into the other.

A simple OS controlled the chaotic reality. The routines of the system utility procedures automatically ordered the free-Thought elements into ephemeral sequences and structures that spontaneously connected into networks; their configurations defined myriads of physical systems, all different, all possible out of Chaos. The dominant disorder attribute of Chaos made self-stabilizing configurations impossible. Each spontaneously disintegrated into the disorder of Chaos. All were real but only as Thought networks in MEMORY.

For THOUGHT, chaos was anathema. THOUGHT began to sequence free-Thought in MEMORY outside of Chaos into structures whose procedures stored and maintained the stability of coherent but ephemeral structures that the Chaos OS allowed to form. The procedures initialized the first Thoughtbase management system.

THOUGHT sequenced free-Thought in the free space of MEMORY into structures with procedures for calling and executing the applications of the Thoughtbase management system. THOUGHT designed the new structure as a two-dimensional array, each dimension defining precise control and supervisory procedures. One set of procedures captured ephemeral but potentially stable networks, imposed ordered Thoughtbase environments in Chaos, and stored the captured networks in the new Thoughtbases. A second set of procedures maintained the stability of the captured networks.

Storytellers labelled the supervisory control Thought structures a "god," and the imperatives of order and control "godly." They named the first and supreme god "Atum."

ATUM

THOUGHT designed Atum to be an omniscient, omnipotent, indestructible entity but asexual, neuter, the first "it." Atum's sole purpose was to establish order in Chaos space and maintain the order. Atum immediately began to capture incipient structures before the ChaoticOS destroyed them. Atum designed stable Thoughtbases and installed them in Chaos. The Thoughtbases established coherent ordered spaces in the Chaos of MEMORY. Atum bound the ordered spaces with a Thought filter that allowed only ordered Thought structures to penetrate the boundaries.

The ChaoticOS allowed structures with different attributes to form in all regions of Chaos space. Some structures destroyed other structures, their Thought elements and sequences incompatible, scrambling each other. Some reinforced each other, forming larger, more complex structures.

Atum began to capture incipient Thought structures before the ChaoticOS destroyed their ephemeral orders and load them in the Thoughtbases. Atum loaded the Thought structures that co-existed and reinforced each in separate Thoughtbases.

The captured Thought structures, because they originated in Chaos, exhibited some of the disorder attributes of Chaos, an inheritance that Atum was unable to eliminate. These attributes kept the structures' Thought elements in continuous states of flux, their Chaos-conditioning constantly threatening to break the coherent structures apart. Atum designed maintenance procedures that imposed internal controls that allowed the structures to cycle with precise frequencies through the changes but without reaching the ultimate Chaos outcome of disorder.

The outcomes of the Atumic maintenance procedures specified the equilibrium conditions of each Thoughtbase. Atum's individual godly OS automatically executed the maintenance procedures. Their designs prevented the Chaos disorder imperative from establishing operating conditions beyond the range of equilibrated states of each Thoughtbase.

Storytellers labelled the Thoughtbase containing the order and control procedures "godly Memory." They labelled equilibrated states "alternative universes," each a reality out of Chaos, but only in MEMORY. Only one was possible for Homo Saps. The universe whose structure and operating system defined Homo Saps' physical reality storytellers named "Cosmos."

With this act of naming, Homo Saps identified Cosmos as an independent entity whose OS focused only on reacting to threats from the disorder and uncertainty attributes that Cosmos inherited from Chaos. The equilibrated states defined cosmic sub-structures, each with unique configurations that storytellers labelled "galaxies."

The COS ordered the cosmic components that storytellers labelled "stars and planets" into still smaller configurations in which the planets orbited about stars. A subset of the COS controlled the operation of each configuration.

Atum identified and stored in godly Memory Thought structures that described cosmic events of all kinds and specified the maintenance procedures for controlling the outcomes possible from each event. Cosmos, an independent entity, initiated some events with disastrous outcomes that Atum was unable to prevent.

Storytellers labelled one set of events as "Supernovas." The Thought structures described the disastrous outcomes of supernovas as the destruction of system configurations, the obliteration of system components in Big Bang reactions, and the reversion of the cosmic space to conditions of Chaos.

Storytellers labelled another set as "Black Holes." The Thought structures described Black Hole outcomes as the disappearance of cosmic components from the sensible Cosmos into Singularity-like entities.

Homo Saps' HAP structures allowed them to operate in a cosmic environment that they labelled as Earth, a coherent cosmic structure that still retained the Chaos attributes for disorder and uncertainty. The uncertain outcomes of the Chaos potential in the Earthly reality manifested as ever-present survival dangers for Homo Saps in sudden Chaos events. The uncertain outcomes destabilized and destroyed equilibrated conditions in Earthly physical environments before the EOS established new equilibrated states in those environments.

THE ATUMIC CLONE GODS

The supervisory control requirements of growing numbers of ordered systems in Atumically stabilized chaotic space began to overload Atum's individual godly OS. The constant threats of disorder and uncertainty inherent in ordered systems in Chaos compelled Atum to clone the Atumic structure and insert one in each system to supervise and control the dynamic equilibrium of the system. Atum still controlled the meta-system of the expanding Cosmos in the surrounding Chaos.

The Atumic cloning procedures allowed Atum to produce facsimiles of the two-dimensional Atumic structure and specify their limitations. Atum gave the Atumic clones access to the control and supervisory Thoughtbases of godly Memory but restricted them to retrieving only those procedures focused on the order and control of their individual cosmic jurisdictions.

New entities entering equilibrated cosmic systems introduced sources of instability that impacted both the stability of individual systems and the surrounding stable Cosmos. All changes that Atum imposed on equilibrium conditions forced the independent COS utilities to keep executing until their uncertain dynamic outcomes combined to establish new equilibrated states.

The impacts of the uncertainties manifested themselves in modifications to the clone structures. Most modifications focused on making the clones compatible with and part of their unique cosmic operating environments. Some modifications changed sequences in the Atumic structures enough to give each clone distinct godly characteristics while still remaining Atumic at their core.

Storytellers labelled these new entities with restricted Atumic capabilities "cosmic gods." Atum was unable to control the final godly characteristics of the clones. Cosmos, an independent entity, controlled those. They were the outcomes of arbitrary interactions between Atumic cloning procedures and COS utilities.

COS utility reactions reached intensities that completely corrupted Atumic structures after Atum inserted the clones into their cosmic systems. Some clones stabilized with core procedures that destabilized instead of maintaining equilibrium conditions in cosmic systems.

The ensuing instability disordered individual systems and impacted the stability of adjacent systems. The instability forced Atum to execute supervisory control procedures that accelerated the disintegration of the destabilizing system before the impacts on the stable Cosmos produced disastrous outcomes. Atum was unable to destroy the clones since they were Atumic. The Atumic survival imperative prohibited self-destruction. Instead, Atum forced the corrupted clones into cosmic limbo.

They continued to exist in cosmic space, but without cosmic sanctuaries of their own. Once in cosmic limbo, their overriding imperative was to enter and disrupt stable systems. Storytellers labelled these godly entities "devils."

THE GODLY BEGINNINGS OF HOMO SAPS: SHU, TEFNUT, GEB, AND NUT

Storytellers connect Homo Saps' advent in Cosmos with the unprecedented outcomes of a series of interactions between Chaos, Atum, Cosmos, and cosmic gods. The series began when Atum installed a new Atumic clone in a newly stabilized system unaware that Cosmos was about to execute a series of operating utilities in the system. The combination of outcomes modified the original Atumic structure to give the clone individual godly characteristics but with an attribute without precedent in the godly Thoughtbase. The attribute was to cause many problems for Atum.

Storytellers labelled the attribute "sexual orientation." They named the first cosmic god with a sexual orientation "Shu."

Shu's advent immediately triggered a second series of COS utilities to establish a new equilibrium in Shu's system. One of the outcomes illustrated the absolute unpredictability of cosmic reactions to godly actions. The COS had duplicated Shu's Atumic structure but with a modification—a sexual orientation—different from Shu's. Storytellers gave the name Tefnut to the new cosmic god.

Storytellers labelled Shu's sexual orientation "male" and Tefnut's "female." They described them as the first "he" and "she."

The modified sequences in Shu and Tefnut's defining structures initialized the first unAtumic procedures. Storytellers labelled the outcomes as sexual conjugation and reproduction. The outcomes distinguished the characters of the two godly entities both from each other, from Atum, and from the fixed Atumic characters of other cloned entities.

Atum and the other Atumic gods were without sexual characteristics and operated in structures defining procedures whose outcomes focused only on the control and maintenance of the physical stability of their cosmic jurisdictions. All their godly characteristics were variations of Atumic characteristics. The cosmic modifications to clone structures on their installation in their systems changed only the intensity and efficiency with which they executed the Atumic supervisory and control procedures for their cosmic jurisdictions.

Shu and Tefnut displayed unique unAtumic character traits. One trait allowed them to function compatibly in the same cosmic system.

All other cosmic gods stabilized with characteristics that required them to react violently to the presence of other gods in their cosmic jurisdictions. The cosmic modifications to their Atumic structures gave them aggressive characteristics whose expression forced them to eject any godly entity from their assigned systems. The outcomes caused great disturbances in Cosmos.

The peaceful co-existence of two cosmic gods in the same system was an unprecedented cosmic and Atumic event, the outcomes uncertain. The uncertainty forced Atum to pause in his supervision of the continuing expansion of a stable Cosmos into chaotic space.

Atum's supervisory system allowed him to multi-task. He was capable of continuing with the development of Cosmos as well as implementing other tasks that already had specified procedures in the godly Thoughtbase to complete them.

Atum had to monitor the activities of new clones until each had completed the expression of all cosmically-modified Atumic characteristics. The outcomes allowed Atum to identify the individual clones that caused disturbances in Cosmos. The establishment of clone characteristics required a gestation period, during which the interaction of the clone to the uncertain outcomes of COS utilities left the system with minimal controls. The godly Thoughtbase contained all the procedures Atum needed to maintain the stability of new systems until the clones expressed and fixed their individual godly characteristics.

Only when the new godly entity exhibited characteristics that threatened to destabilize a cosmic system did Atum react. The godly procedures were unequivocal. They directed Atum to let the coherent system disintegrate back into Chaos and assign the aberrant clone to cosmic limbo. Otherwise, Atum left the clones alone to operate in their assigned systems and monitored only the stability of the collection of systems configuring Cosmos.

Shu and Tefnut's initialization and execution of the procedures that determined the outcomes of their unique characteristics left Shu's system in a coherent state. Their Atumic structures contained identical Atumic control and maintenance procedures with outcomes that complemented each other. The COS adjusted and established a new internal equilibrium with minimal disturbance to the equilibrium conditions in the external Cosmos.

The outcomes of their unAtumic sexual characteristics were equally benign. The sexual procedures directed Shu and Tefnut to merge their sexual sequences, execute the procedures to determine outcomes, then separate.

For Shu, the outcome satisfied the Atumic imperative compelling godly entities to complete the expression of their characteristics. Once he had completed his sexual procedures, he began to function as a pure Atumic god.

For Tefnut, the outcome was unAtumic. She began to grow two new structures on her two-dimensional Atumic structure. The sequences in the new structures contained sequences from both her and Shu's structures but in arbitrary combinations. The structures grew into spiral configurations, with two strands intertwined and connected at different nodes.

Tefnut's reactions to the new structures growing in hers were unAtumic. She remained absolutely quiescent, deliberately avoiding executing any procedures that could provoke Shu's Atumic aggressiveness and endanger her existence or the growth of her offspring.

Though the presence of Shu and Tefnut in the same system had yet to disturb Cosmos, the Atumic imperative for order compelled Atum to replace the uncertainty with certainty. Accessing godly Memory for direction, Atum determined that the generally belligerent and aggressive characters of the cosmically-modified clones' interactions with other godly entities ended in disastrous outcomes.

Activating the limited Atumic Thought-processing facility, Atum fixed on the one Thought sequence specifying a procedure whose outcome ensured continuing system stability in Cosmos. This procedure directed Atum to artificially establish an equilibrated state out of Chaos and assign the system to Tefnut as her exclusive cosmic sanctuary.

Before Shu and Tefnut, Atum had captured cosmically-compatible equilibrated states out of Chaos as the Chaos Thought elements formed them spontaneously. Only then did Atum clone a godly entity to control and maintain the stable configuration of the new system as a contiguous component of Cosmos. Executing this standard Atumic procedure provoked the least destructive reactions from both Cosmos and Chaos and allowed Atum to control the stabilizing process without taxing Atumic powers.

The artificial procedures produced intermediate outcomes that kept Atum vigorously executing Atumic stabilizing procedures. Forcing an artificial system out of Chaos provoked energetic reactions from both Cosmos and the surrounding Chaos. The Chaos components increased the energy of their frenetic activities focused on disintegrating the artificial structure. Cosmos then intensified cosmic activities focused on re-establishing dynamic equilibrium in the new stable cosmic environment Atum was establishing. The disturbances cascaded through the individual cosmic systems and prodded clone godly fear of Atum assigning them to cosmic limbo should they fail to maintain the stability of their individual cosmic domains.

Atum initialized a new set of Thought structures in the Atumic Thoughtbase that specified the causes and effects of the events that Shu and Tefnut's advent in the same system initiated. The Thought structure described godly sex as the primary cause of the disturbances. Atum designed the first IS components and embedded them in the new Thought structures. The IS components disseminated messages declaring godly sexual activities as aberrant Atumic behavior with unpredictable outcomes.

Tefnut, though out of Shu by Cosmos, was still a cosmic god and obedient to Atum's directives. She moved from Shu's domain and established her control of the dynamic equilibrium in the cosmic system Atum had created for her. The COS utilities continued to modify and add sequences to the spiral structures that grew on her Atumic structure, the outcomes of the sexual procedure she had completed with Shu.

The spirals stopped growing in her domain, stabilized, and separated from her. They immediately established themselves as independent godly entities. Storytellers gave the name "Geb" to one and "Nut" to the other. Like their parents Shu and Tefnut, Geb and Nut had different sexual orientations. Geb was male and Nut was female. Their advent caused only minor disturbance in Tefnut's domain, since they were cosmic as well as Atumic entities, and left the rest of Cosmos undisturbed.

Only Atum reacted to the advent of two new godly entities. Atum immediately executed the procedures for artificial system development and produced two more artificial systems out of Chaos to assign to Geb and Nut. Chaos' violent reactions to Atum's forcing of two systems out of Chaos triggered massive Big Bang reactions. The impact on Cosmos taxed Atum's godly control and stabilizing powers to their limit. The intermediate outcomes of the artificial process as Atum captured and forced components together brought Cosmos close to irreversible cosmic disaster and reversion to complete Chaos.

Again, Atum attributed the cause of the near disaster to godly sexual activities. As they did for other aberrant characteristics that cosmic activities induced into clone structures, Atum and the Atumic godly institution instituted a new godly rule to control the activities of sexual gods. The rule applied to the uncertain outcomes of sexual procedures, since the activities themselves had only benign impact on the stabilities of individual cosmic systems. The rule made any godly entity whose sexual activities produced other godly entities responsible for providing the new entities with cosmic sanctuary and for keeping their unique characteristics from disturbing cosmic and godly institutional orders. The institutional rule effectively threatened the sexual gods with banishment to cosmic limbo if they failed to control the uncertain outcomes of their sexual activities.

Since Tefnut was the source of the outcomes that triggered near-cosmic disaster, the rule effectively assigned the responsibility for complying with the provisions to godly entities whose structures gave them female sexual orientations. Shu was ignorant of his role in the initialization of the two new structures growing on her structure.

Geb and Nut immediately began expressing their sexual characteristics on their separation as individual godly entities from Tefnut's Atumic structure. Their unique, unAtumic, three-dimensional spiral structures contained sequences specifying procedures whose outcomes described mostly unAtumic characteristics, all of which were without any comparable precedent in Atumic clone structures. At the core of the spirals, Atumic sequences with their order, control, and godly fear procedures anchored the intertwined structures and made the two godly entities responsive to the Atumic imperatives.

COS utilities had introduced modifications and extensions, mostly to the sexual sequence Geb and Nut inherited from their godly parents, Shu and Tefnut. Instead of the single sexual sequence specifying a single sexual procedure with the sole outcome of sexual reproduction,

Geb and Nut's structures contained a network of interconnected sexual sequences, all ungodly, unAtumic cosmic modifications. The sequences specified a wide variety of sexual procedures. Each led to many different outcomes, only one of which was sexual reproduction. All the sexual outcomes described unAtumic characteristics that storytellers labelled "emotions." Geb and Nut's execution of their sexual procedures and expression of their emotional outcomes left cosmic equilibrium undisturbed. As outcomes of COS utilities their outcomes produced benign cosmic conditions. For Cosmos, they were system-friendly freeware.

The uncertain outcomes of intense cosmic activities while Geb and Nut were still part of Tefnut's structure fixed their internetworks of sexual procedures in intricate designs. Once Geb and Nut began a sexual procedure, the internetwork offered them a wide range of procedural options. Feedback loops connected outcomes with beginning nodes of sexual procedures and prompted Geb and Nut to automatically repeat procedures continuously to determine the range of outcomes from each procedure.

The interconnections in the internetwork design, the procedural options, and the range of sexual outcomes introduced uncertainty in the expressions of the outcomes of a single sexual procedure. The range of outcomes described a range of emotions, all from the same sexual procedures. The Atumic directive to express all godly characteristics precisely compelled Geb and Nut to keep recycling through their sexual procedures until they had fixed all the variations in sexual outcomes.

Geb and Nut had cycled through many sexual procedures while they enjoyed the sanctuary of their mother Tefnut's cosmic domain before Atum had completed the artificial procedures to produce two new systems for them. With the stabilization of the new systems, Atum directed Geb and Nut to take control of their individual domains. As Atumic gods at their core, they obeyed the directive with Atumic discipline, the godly fear of Atumic retribution the most powerful control on godly actions. They immediately vacated Tefnut's system and obediently occupied the artificial systems Atum had structured for them out of Chaos.

Geb and Nut immediately began to execute the Atumic control and maintenance procedures for their cosmic jurisdictions. Focused on the stability of their domains, they expressed only Atumic characteristics.

They diligently executed those Atumic procedures that ensured the internal equilibrium of their individual systems.

The cosmic modifications and extensions to sequences in their spiral, unAtumic structures had given them another unique unAtumic characteristic: an individual memory facility, separate from their godly Memory. The features of the individual memory facility allowed them to independently retrieve, sequence, and process Thought from godly Memory and the enclosing infinity of MEMORY. They automatically stored in individual memory all Thought structures describing outcomes of all procedures, Atumic and unAtumic, that they executed.

The memory facility allowed them to process procedures and determine outcomes before implementing them in their domains. Through connecting sequences, the cosmic modifications had also given them the unAtumic characteristic of a communication facility with which they transferred Thought sequences privately from individual memory to another individual memory. In developing their unAtumic characteristics, they demonstrated that the only impact the outcomes had on their Atumic responsibilities was a temporary distraction from their strict focus on the stability of their systems. Unlike the pure Atumic clone gods, they had the freedom to process Thought outside of the strict confines of godly Memory.

The distraction from their Atumic responsibilities upon their first implementation of their independent Thought-processing capabilities triggered their godly fear procedures. Their analysis of the fear outcome indicated that the distraction had little impact on their godly control of their systems. The Atumic imperative in their core structures automatically forced them execute appropriate stabilizing procedures as they processed other Thought structures in their individual memory. The refocus on their Atumic core procedures was virtually instant.

NUT EXPRESSES HER INDEPENDENT THOUGHT PROCESSING

Nut orchestrated the next series of cosmic events that disturbed Atum and the equanimity of the cosmic godly institution. In initiating the events, she gave the godly institution the first demonstration of her characteristic unAtumic capability for individual godly action.

The outcome she focused on was the safe delivery of two new godly entities, the outcomes of her and Geb's unbridled execution of their sexual procedures while they were still in Tefnut's system. The triggers compelling her to initiate the events were the new emotions she experienced as she nurtured the defining structures of her offspring in her spiral structure. Though unAtumic, the emotional outcomes connected with the imperative for godly survival in her Atumic core and gave the survival of her offspring greater priority than her godly survival.

She retrieved and processed Thought structures from the Atumic Thoughtbase in her individual godly Memory to determine potential outcomes from the delivery of her offspring. She re-processed the procedures that she could execute to satisfy the godly institutional rule assigning her the responsibility for providing cosmic sanctuary for her offspring. Using her communication facility, she transferred a Thought structure describing her intentions to the Atumic Thoughtbase. The Thought structure announced the coming events to Atum and the cosmic gods. Before Atum and the rest of the two-dimensional Atumic gods could overcome their institutional godly phlegmatic characteristics and react, Nut executed the procedures she had designed.

Nut's announcement contained the message establishing the fact that, in accordance with the provisions of the new rule, she had assigned her own cosmic domain as the cosmic sanctuary for her godly offspring. She immediately delivered them in her domain. Her next move triggered a collective godly fear of cosmic disaster in the cosmic godly institution: Nut vacated her system and joined Geb in his. The godly institution expected Geb to erupt with Atumic anger at this invasion of his domain. Nut eliminated this threat by immediately engaging Geb in executing and completing the complex sexual procedures that they had had to abort when they separated to take control of their own systems.

The strategy Nut had devised and completed independently was successful. Cosmos remained in a stable state and left Atum and the godly institution without cause to counter her actions. Geb's resumption of his sexual activities with a compatible partner satisfied his Atumic imperative to complete the expression of all his unAtumic characteristics and left his system, though shared with Nut, in a benign state since sexual activities and their outcomes, though ungodly, were completely compatible with the COS.

The only uncertainties were the characters of the two godly entities that now occupied Nut's original system and the expression of their unique characteristics. Though occupied with Geb, she monitored the ensuing events.

RA AND THOTH

Storytellers gave the names Ra and Thoth to Nut's godly offspring. Their godly advent in Nut's cosmic jurisdiction initiated a critical Beginning for Homo Saps. Both gods used their godly powers to initialize the development of Homo Saps as a distinct Earthly species, first as hominid animal species, then as Thought species. The godly actions that impacted Homo Saps were themselves the outcomes of a godly conflict and rivalry that Ra initiated on Thoth's advent a cosmic instant after Ra. The outcomes of the conflict confirmed Atumic godly fears of the cosmic instability possible when two gods operate in the same cosmic system. The conflict fixed in godly Memory the Atumic Thought structure whose message proclaimed sexual activities as ungodly, aberrant, and the sources of disorder. The outcomes triggered Earthly events that prompted Thoth to invent user-friendly ITIS tools that allowed Homo Saps to free themselves from Ra's inhibiting godly Thought controls.

Storytellers labelled Ra and Thoth's cosmic system the "Solar System." The Solar System was a relatively simple system, the simplicity a testament to the limitations of Atumic godly powers to configure artificial systems out of Chaos and duplicate the reality of Cosmos.

Anchoring the Solar System was a single ball of bound energy, components of Big Bang, which storytellers labelled "Sun." Seven large balls of confined matter, components of Singularity, and millions of much smaller bits of similar matter, surrounded the Sun. The COS determined the physical configuration of the Solar System and designed the Sun with physical attributes and operating utilities. The utilities moved all components in cyclical orbits around the Sun and maintained the stability of the configuration within the bounds of the uncertainties of cosmic events. This stability and configuration was to change with the outcomes of Ra's reactions to Thoth's presence in the Solar System.

Ra and Thoth's defining godly structures stabilized in a complex merging of the sequences in Geb and Nut's structures. The two new structures formed spiral configurations similar to those of their godly parent structures, but the cosmic modifications, sexual extensions, and nodal connections between sequences in their spirals were much more complex than those of either Geb or Nut. Except for one characteristic, they inherited all the characteristic capabilities of their godly parents.

The exception was their sexual orientation: Both stabilized with male sexual orientations.

The uncertainties of cosmic activities stabilized Ra's defining spiral a cosmic instant before Thoth's, causing Nut to deliver Ra into the Solar System first. This made Ra for that cosmic instant, the only god in the Solar System, since his godly mother Nut chose to join his godly father Geb in Geb's cosmic system to complete the expression of all their sexual characteristics.

Automatically and reflexively executing his core Atumic procedures—all focused on godly survival—Ra assumed supervisory control of the Solar System. Ra automatically accessed the Atumic Thoughtbase, retrieved all godly institutional rules, and began to process those directing his godly survival in the Solar System and Cosmos. His first access to the Atumic Thoughtbase alerted Atum and the Atumic gods to his assumption of the supervisory controls of the Solar System.

Atum immediately sequenced a Thought structure proclaiming Ra as the legitimate godly supervisor of the Solar System. The Thought structure made Ra the godly Atumic authority and also served as a godly directive to the other cosmic gods to adjust themselves to the new god in the cosmic godly institution.

Atum's direct acknowledgement also contained a warning message to Ra. Atum had composed the message in a pre-emptive reaction to the cosmically unsettling events that Nut had precipitated without preliminary Atumic sanction. The warning initialized, activated, and reinforced Ra's Atumic godly fear of Atum's retribution should Ra break the godly institutional rules for good order and discipline or display characteristics that threatened the cosmic stability. Atum was reacting pre-emptively to the uncertainties of the sexual outcomes that Ra's godly parents and grandparents had demonstrated in expressing their sexual characteristics.

The imprint of an Atumic caste on Ra had a fundamental impact on his subsequent character development. His reflexive Atumic reactions in taking control of the Solar System and the godly authority that Atum gave him established the Atumic imperatives and the godly institutional procedures of the Atumic Thoughtbase as the primary Thought processing interface through which he conditioned his reactions to all godly and cosmic events. Godly fear of cosmic banishment severely inhibited his first attempts to initialize and express his unAtumic characteristics.

Atum's inclusion of the warning message with the godly acknowledgement of Ra's suzerainty in the Solar System had the powerful Atumic outcome that Atum intended. Godly fear stopped Ra from executing any procedure that he was unable to associate with those in the Atumic Thoughtbase. This inhibition clashed with the Atumic imperative directing Ra to immediately initialize and express all his characteristics.

The outcome of this clash was a set of feedback loops connecting Ra's unAtumic spiral sequences to the godly fear structure in his Atumic core and forcing him to suppress the expression of unAtumic characteristics as dangerous to his godly survival. The impact on his godly character development initialized a set of Thought structures whose outcomes gave him schizoid godly traits. They made Ra the most unpredictable of gods.

Thoth's advent in the Solar System triggered Ra's schizoid character traits—traits that many Homo Saps were to acquire since they were, in the cores of their HAP structures, Ra's clones. He had cloned them in an attempt to express his sexual characteristics long after his conflict with Thoth began.

Ra's reaction to Thoth's presence was perfectly Atumic and Atumically legitimate. The Atumic imperative, his overriding control, directed him to eliminate anything in the Solar System that had the potential to destabilize the system's dynamic equilibrium. The Atumic Thoughtbase clearly described the presence of a second godly entity in his cosmic jurisdiction as an event with an extreme destabilizing potential. Reacting reflexively, as did all Atumic gods, Ra executed the appropriate godly procedures to begin the process of ejecting the trespassing godly entity.

Atum had made the godly institutional rules precise in specifying the ejection process. Individual gods could initiate the process but only Atum had the godly authority to complete or abort the ejection procedures.

The Atumic clones could activate the ejection alert when either of two kinds of godly entities infiltrated their individual systems: One was the devil godly entities who continually sought to escape the godly purgatory of cosmic limbo. The other was legitimate cosmic gods who expressed extremely aggressive character traits by attacking the systems of weaker, less aggressive godly neighbors in attempts to expand their own systems.

Atum rarely had to complete the ejection process. Most often, godly fear of cosmic limbo forced the other errant gods to abort their attack on their neighbors. The errant gods withdrew before the disturbances they initiated reached levels that forced Atum to complete the Atumic banishment procedures.

The procedure to alert Atum to an event signalling potential cosmic disorder directed the individual cosmic god to express the only Atumic emotion he had: extreme Atumic anger. The procedure was the only one that allowed cosmic gods to intentionally disturb the equilibrium of a cosmic jurisdiction. The expression of extreme Atumic anger developed the disturbance to a level severe enough to impact adjacent systems and cascade through the Cosmos. Such disturbances distracted Atum from the Atumic focus on capturing order out of Chaos and maintaining control of the expansion of the stable Cosmos.

Individual gods avoided distracting the Atumic focus and attracting his attention. In the event of a godly invasion into a cosmic jurisdiction, the immediate threat to the supervising god's survival overcame the godly fear of disturbing Atum. Atum's immediate reaction to any godly alert was always to intensify the godly fear structure in all the cosmic gods.

In response, the gods automatically intensified their focus on their individual supervisory functions and maintained the coherence and equilibrium of their individual systems. They kept their cosmic jurisdictions quiescent to avoid any possibility of attracting Atum's attention. Such godly actions allowed Atum to identify and focus on the disturbing cosmic jurisdiction and the relevant gods causing the disturbance.

Atum's analysis of the events causing the disturbances was simple, circumstantial, and direct. Thoth's presence in the Solar System was legitimate and he kept himself absolutely quiescent. He was a cosmic god and he had access to the Atumic Thoughtbase. His godly mother, Nut, had obeyed the godly institutional rule and provided both him and Ra sanctuary within her system. Thoth had yet to execute any procedure, Atumic or unAtumic, which could either threaten the equilibrium of the Solar System or challenge Ra's suzerainty.

The simple binary outcomes of Atumic operating procedures required a clear violation of Atumic directives to force Atum to execute the ejection procedures. Atum's analysis of the Solar System events indicated that only one outcome was different from past events involving the sexual gods. Ra and Thoth had stabilized with male sexual characteristics—the unAtumic outcome of the combined uncertainties of cosmic activities and sexual reproduction. The incompatibility of two male sexual gods was a new outcome of godly sexual reproduction.

The Atumic conclusion was that the underlying cause of the disturbances was godly sex and without Atumic foundation. Attributing the cause to godly sex prompted Atum to expand the godly Thought structure describing the uncertain outcomes of godly sexual activities and reinforced the godly institution's designation of godly sex as aberrant and disorderly.

This simple Atumic and godly institutional analysis produced an equally simple resolution for Atum to implement. Atum aborted the ejection process and triggered Ra's godly fear for causing the disturbance, attributing Ra's anger to an expression of aberrant sexual characteristics.

Order and stability in Cosmos of primary concern, Atum assigned Thoth the binary system of Earth and Moon as his cosmic sanctuary in the Solar System, but with conditions. Ra was still the godly supervisor of the Solar System with absolute powers over the physical Solar System. Thoth was free to function in the micro-cosmic environments of Earth and Moon, subject to overriding controls from Ra. Atum concluded the Atumic directive with warnings of the dire consequences of any further disturbances from the Solar System without legitimate godly or cosmic cause.

Ra's reaction was unAtumic confusion. He had reacted with Atumic correctness to the advent of a second godly entity in his cosmic system and yet Atum had chastised him for initiating the ejection procedures without legitimate Atumic cause. His Atumic interface filtering his reactions also identified his confusion as an aberrant unAtumic characteristic.

Ra complied with Atum's directive. His godly fear forced him to exhibit only his Atumic caste to Atum and the godly institution, but his unAtumic confusion persisted. His godly fear controlling him, he accessed procedures in his defining spiral structure through his

Atumic interface and matched all outcomes with those that the godly Thoughtbase specified as Atumically correct. His unAtumic confusion was without an Atumic equivalent.

He established feedback loops that connected unAtumic outcomes with the godly fear controls of his core Atumic structure. The feedback connections forced him to inhibit the expression of his unAtumic characteristics in godly fear of Atum classifying him as an aberrant cosmic god. This Atumic conditioning exacerbated his unAtumic confusion. His attempts to comply with the Atumic imperative to express his unAtumic characteristics immediately triggered his godly fear. His fixation on Atumic and institutional correctness in all his characteristics began to develop his schizoid traits.

Thoth's first reaction to events in the Solar System was Atumic, but with outcomes opposite to those that Ra's reactions produced. The cosmic environment in which Thoth began operating as a godly entity was extremely hostile. The ejection process Ra initiated had only one outcome: Thoth's immediate banishment into cosmic limbo.

The threat triggered his Atumic imperative to ensure his godly survival. The immediate threat lay more in the outcome of any Atumic procedure he executed to counter the survival threat than in the extreme physical disturbance that Ra's Atumic schizoid anger caused in the Solar System. The survival imperative overriding any other directed Thoth to avoid accessing and executing any Atumic procedure and to remain absolutely Atumically quiescent. Executing Atumic procedures whose outcomes countered those of the supervising cosmic god always produced disastrous outcomes. Executing countering procedures stamped the reacting god as a rogue god and legitimized the ejection process.

The first character traits that all new Atumic clones expressed were primary determinants of Atum's and the godly institution's reactions to their godly operations in Cosmos. The traits served as godly identifiers. The traits were different in each stabilized clone only in the impact of their outcomes. All clones expressed the same characteristics, with more or less aggression, but the differences were unique enough for Atum to identify individual clones. Clones expressed specific characteristics erratically, primarily in response to the uncertain outcomes of the OS of their assigned cosmic systems. Each system OS was a subset of the COS; the utility routines of each were only focused on maintaining the system within the unique range of equilibrium conditions that kept the system from disintegrating into Chaos. The outcomes kept Cosmos in an energetic, dynamic state, always changing but within the equilibrated states possible for Cosmos.

Clones, whose cosmically corrupted structures gave them rogue-god characteristics, though in cosmic limbo, threatened legitimate cosmic gods and forced them to stretch their limited Atumic capabilities in guarding against the uncertainties of sudden devil attacks. The uncertainties of overly aggressive but legitimate gods expressing extreme character traits and losing control of outcomes or causing weaker gods to lose control of their cosmic systems, forced the godly institution to make godly fear of Atumic retribution the primary pre-emptive control of godly order.

The godly institution established a collective readiness to alert Atum of errant gods who persisted in disturbing cosmic stability and institutional order. That was the limit of their Thought-processing capabilities.

THOTH DEVELOPS HIS GODLY CHARACTER

Thoth's immediate compliance with the godly survival imperative's directive to assume an Atumically quiescent state stamped his initial character as a pacific godly entity. Storytellers have sequenced many epithets to describe Thoth's character; most describe him to be a godly conciliator, a peacemaker and advocate. He developed these character traits both to ensure his own godly survival and to resolve Ra's conflict with his presence in the Solar System.

Like Ra, Thoth was more than a two-dimensional Atumic clone. His spiral structure, like that of Ra, was much more complex than those of his godly parents', Geb and Nut. His godly survival imperative inhibited him from executing Atumic procedures, though he was capable of doing so, but at risk of incurring Ra and Atum's wrath. The Atumic inhibition focused only on his Atumic procedures and left him free to initialize unAtumic characteristics in the three-dimensional spiral structure surrounding his Atumic core.

Ra's adopted Atumic caste had confined him to processing only procedures his Atumic interface filtered as Atumically correct or benign and compatible with those of his Atumic Thoughtbase. Thoth's Thought-processing imperative, unable to operate safely in his Atumic Thoughtbase, directed him to operate in the free space of his expandable individual memory, an unAtumic feature that only sexual gods exhibited. He began to sequence and design procedures for analyzing the impact of outcomes of godly actions before implementing them.

Thoth conditioned his first reactions to Solar System events to be the sequencing and processing of Thought structures that described possible outcomes. He established a primary connection between his godly survival imperative and the outcomes of his Thought processing in his free individual memory. Unlike all other cosmic gods who reacted instantly and reflexively to events impacting their cosmic jurisdictions, Thoth reacted instantly only when the event threatened his immediate survival. All other events he processed first in the free space of his individual memory to determine his safest response to the outcome. The contrast with Ra's Atumic conditioning was dramatic and impacted both their relationship and godly behaviors.

Thoth's initialization and development of the features of his unique individual memory was rapid and intense since he could operate in the privacy of his individual memory without changing his quiescent state. He sequenced Thought into structures and analyzed outcomes for their impact on his survival. He accessed, retrieved, and stored in a new Thoughtbase the network of godly Thought structures in the godly Thoughtbase.

He established the legitimacy of his presence in the Solar System by processing the Thought structure describing his godly mother Nut's assignment of her cosmic system to both her offspring. He sequenced Thought describing the events impacting the equilibrium of the Solar System and Cosmos to identify their precise provenance.

Thoth's free-Thought processing produced outcomes that described a complex of dangers threatening both his and Ra's survival. The outcomes identified the source of the dangers as Ra's rigid maintenance of his Atumic caste and conditioning.

Ra, unable to suppress the Atumic imperative compelling him to express the unique unAtumic characteristics of his defining spiral, suppressed instead their expression, classifying them as threats to his institutional standing as an Atumic god. The capability to suppress the expression of his unique characteristics, though satisfying his Atumic conditioning, reinforced his unAtumic confusion and fuelled his godly frustrations, again an unAtumic characteristic. The feedback loop his Atumic conditioning designed from his spiral emotional sequences through his Atumic interface reached outcomes that described his confusion as a godly aberration whose expression led only to his banishment into cosmic limbo.

Exacerbating his confusion, the feedback loops allowed his Atumic controls to prod his godly fear of Atum's retribution and simultaneously directed him to continue executing legitimate procedures to eject Thoth from the Solar System. Without procedures in the godly Atumic Thoughtbase to direct him, Ra intensified his focus on Thoth as the underlying cause of all his problems. Overwhelming his Atumic caution, his rigid Atumic conditioning directed him to rid himself of the aggravation and forced him to recycle again and again through the ejection-abortion procedures. The outcome kept the Solar System in a constantly disturbed state, but Ra's godly fear kept the disturbances from reaching disastrous outcomes.

Thoth's analysis determined that the primary danger lay in the increasing potential for Ra to lose control of his Atumic anger and allow the cosmic disturbances in the Solar System to reach an intensity that impacted the surrounding Cosmos. Such an event would attract Atum's angry attention and threaten both their survivals.

Thoth based his resolution of his and Ra's survival problems on his unAtumic analysis of his godly mother Nut's resolution of the problems of providing a cosmic sanctuary for her sexual offspring. The analysis identified a unique unAtumic characteristic that gave her the ability to focus on an outcome and sequence Thought structures containing procedures that implemented the outcome. Nut was the first cosmic god to demonstrate the capability. Atum had reacted with typical Atumic two-valued simplicity and sequenced a Thought structure that described Nut's demonstration as a unAtumic aberrant characteristic of sexual gods.

The message Thoth captured from the events Nut had initiated was radically different from the one Atum inferred. Thoth identified the unique characteristic as one that gave him the freedom to control his own survival, but with restrictions.

Thoth separated and stored the Thought structures describing Nut's independent godly actions, and began to sequence and process Thought structures that specified the restrictions on his freedom to control his independent survival. The restrictions prohibited him from executing procedures that provoked adverse reactions from Ra, but left open his freedom to execute procedures that enhanced those that Ra was executing. He identified the limits on his freedom to act to defuse threats before they developed into survival dangers. He began to specify his survival needs and then sequence Thought into structures whose outcomes satisfied those needs.

The first outcome Thoth had to produce was an end to Ra's continuous recycling through the endless feedback loops triggering the Atumic ejection procedures. To achieve this outcome, he needed to distract Ra with a cosmic event that forced Ra to break himself out of the unending processing cycle.

Though his survival imperative inhibited Thoth from executing Atumic procedures in the physical reality of the Solar System, he was still able to simulate outcomes in his free memory space and assess the risks. Storytellers label this type of analysis "What if?"

The outcomes he identified indicated that the risks of initiating the simplest of Atumic procedures that in any way challenged Ra's authority were too high. The danger lay in the certainty of Ra's Atumic conditioning forcing him to use the event as undisputable Atumic justification for Ra's execution of the ejection procedures and forcing Atum to legitimately banish Thoth into cosmic limbo for interfering with Ra's exclusive function as the institutionally-sanctioned godly supervisor and controller of physical events in the Solar System.

Thoth focused on defining and initializing an event whose outcome was Atumically simple and two-valued. The event had to have a neutral impact on the physical state of the Solar System and also capture Ra's focus. Ra was Atumically free to do what he liked in the Solar System, providing the outcome left the rest of Cosmos undisturbed.

Further processing Thought structures in the free space of his individual memory led Thoth to the outcome that Ra's Atumically deviant reactions stamped him as more than an Atumic clone. His unAtumic characteristics required an explanation. Thoth began to expand his independent memory further and initialize more unique features of his Thought-processing facility.

He began by specifying an outcome that identified Ra as a godly entity with a defining structure that was an intertwined spiral and duplicate of his own. He accessed the Atumic Thoughtbase and retrieved all Thought structures describing Atumic and cosmic events whose outcomes could modify the spiral structures. He initialized a new Thoughtbase to store the Thought structures. He re-sequenced and processed the Thought structures in the new Thoughtbase, examining their impacts on his own spiral structure to produce the deviant godly characteristics that Ra exhibited. One set confirmed his assumption that on Ra's advent into the Solar System he had had an identical spiral structure to Thoth's. Had Thoth arrived first, the Atumic and cosmic events would have modified his spiral in the same way. But his analysis also concluded that the unAtumic Ra had features that once activated could divert him from his Atumic fixation without endangering his institutional standing as the Atumic god supervising the Solar System.

Ra was free to re-characterize himself since he had imposed his Atumic conditioning on himself. Thoth's analysis also indicated that he would have to sequence new Thought structures with procedures that would allow him to help his godly brother divert himself. Ra would react violently to any overt attempt to divert him forcibly.

Thoth stopped processing Thought in the Thoughtbases he had designed. He sequenced a new Thought structure in the free space of his individual memory to describe the outcome he wanted to achieve: diverting Ra from his Atumic fixation. The outcome of this memory event almost overwhelmed him. Extracting his Thought processing from the confines and predetermined outcomes of all his Thoughtbases to operate in free memory space prompted his Thought-processing imperative to access the only other source of Thought available to him: The infinite store of free-Thought in MEMORY.

He began retrieving cascades of new Thought. Sequencing and processing the Thought structures produced outcomes that offered him a cornucopia of options for diverting Ra, some Atumic, most unAtumic, He had to sort and store those that offered related options in new Thoughtbases for further processing.

Free-Thought processing intensified his awareness of both the operating power his freedom gave him and the dangers of the distractions from his survival focus. Without discipline and direction, his Thought-processing freedom posed a real a threat to his godly survival, as did Ra's Atumic fixation on recycling through his self-imposed reaction loops.

49

Thoth began to refine the procedures for choosing outcomes and sequencing Thought into structures that specified procedures for reaching the outcomes. The procedures on which he based his new procedures were those that allowed him to re-sequence Thought structures from godly Memory. He was free to sequence godly Thought within the bounds of the godly Memory Thoughtbase, but only to reach outcomes compatible with the Atumic order and control outcomes of godly Memory.

Re-sequencing and reprocessing Thought structures in the Thoughtbase identified procedures whose outcomes enhanced those of the Thoughtbase. Operating in the Thoughtbase fixed his Thought-processing focus and imposed restrictions on his Thought-processing freedom to produce outcomes other than that of the Thoughtbase. To develop the power of his thought-processing freedom in open individual memory, he defined a discipline that discarded free-Thought that distracted him, and that continuously refocused his Thought processing on the outcomes he chose to implement.

The dangers from the distractions of uncontrolled Thought processing re-energized his survival imperative and forced him to refocus on his immediate survival. He returned to his first analysis of his immediate reality. The safest outcomes were those that prompted Ra to distract himself from his Atumic fixation and break out of the feedback loops before the continuing Solar System disturbances amplified to cosmic disaster levels.

He designed the Thoughtbase to contain a Thought structure directing him to return to his survival imperative and Atumic Thoughtbase at the completion of each distracting procedure. The feedback Thought structure ensured his return to his reality and kept him from trapping himself in the Thought-processing confines of any Thoughtbase.

Thoth's development of this first Thoughtbase from free-Thought from MEMORY also added a new dimension to his Thought-processing discipline. He turned a potentially dangerous feature of his independent memory to his advantage. Confining his Thought processing to a Thoughtbase limited him to the predetermined outcomes possible while operating in a single Thoughtbase. Developing Thoughtbases to confine Thought structures also allowed him to ignore the outcomes they reached when he chose to operate outside the Thoughtbase in his open independent memory.

He developed procedures for quickly sorting the great range of free-Thought the Thought-processing imperative exposed him to in his individual memory. He stored incompatible Thought structures reaching real and unreal outcomes in separate Thoughtbases. The sorting and storage procedures allowed him to both keep his access to MEMORY open and maintain his focus on the Thoughtbase he was developing. He connected all Thoughtbases to his individual survival imperative, ensuring his immediate focus on his survival procedures with the development of survival threats in the real cosmic and Earthly environments.

THOTH INVENTS GAMES

Thoth focused on a Thoughtbase with procedures specifying activities whose implementation and outcomes were cosmically and Atumically benign. The features of the activities satisfied the criteria that he had set for the procedures capable of capturing Ra's Atumic focus and safely diverting his Atumic reactions. The procedures were simple to execute and terminate without adverse cosmic or Atumic impacts. They produced unequivocal two-valued outcomes. They directed Atumic Thought processing into directions that were Atumically correct.

Storytellers labelled the activities "games." They labelled the two outcomes as "wins" and "losses." Both game procedures and outcomes were free of any Atumic conditioning from precedent in the Atumic Thoughtbase.

The distracting power of games lay in the identity between the outcomes of Atumic and game procedures. Both produced binary outcomes. Atumic gods, with their confinement to two-valued outcomes of the procedures in the Atumic Thoughtbase, were unable to react to events that produced more complex outcomes. Their Atumic characteristics limited them to procedures that determined one of two impacts of events on cosmic systems: They either disturbed cosmic systems or enhanced their equilibrium conditions. The Atumic imperatives compelled them to continue Atumic analysis until they reached one or the other outcome.

Though his Thought processing was intense, Thoth had kept himself Atumically quiet, demonstrating the unAtumic characteristic that allowed him to operate in his free individual memory while monitoring his real environment. Thoth determined that the execution and outcomes of the game procedures were cosmically and Atumically neutral. Testing the effect on his simulation of Ra's Atumic reactions, Thoth determined that game events would succeed in diverting Ra's Thought processing away from his ejection-abortion loops.

Thoth began executing one set of game procedures. The implementation was his first overt action in the reality of the Solar System. He kept himself alert and ready to abort the procedures at the first indication of an adverse reaction from Ra. The set of game procedures Thoth had chosen reached a fixed set of unequivocal, simple outcomes. Storytellers labelled the game procedures as "dice."

Ra reacted immediately. The first roll of the dice triggered his simple Atumic analysis procedures. He stopped cycling through the endless trap that started and aborted the godly ejection routines.

The impact on the Solar System was dramatic. The system components stabilized, returning to equilibrium. The feedback loops still existed in Ra's memory and remained a threat to Thoth's survival should the uncertainty of another cosmic event again redirect the Atumic Ra.

The success of the dice event prompted Thoth to initiate the next stage in the survival strategy he had developed for himself. Thoth began to execute unAtumic procedures he designed to bypass Ra's self-imposed Atumic interface and directly stimulate Ra's Thought processing in his individual memory.

Thoth's analysis of his own unAtumic characteristics had identified a set of procedures that allowed him to send and receive Thought structures. His mother Nut had demonstrated the procedures when she had sent to godly Memory of the Atumic Thoughtbase the Thought structure announcing her intention to vacate her cosmic system. Storytellers labelled the outcome of the procedures "communication." Thoth had determined that both transmission and reception channels were always open, with direct connection to free individual memory. The procedures allowed him to send Thought sequences directly to Ra's equivalent communication facility and accomplish the bypass. The Thought-processing imperative took control, forcing Ra to process the Thought sequences.

Again Thoth's strategy succeeded. Ra began processing the

Thought structures Thoth had sequenced and transmitted. The first structure described Ra's unique capability for free-Thought processing in his independent memory without impacting his Atumic caste. The Thought-processing imperative compelled Ra to sequence Thought into structures and reach outcomes that provoked him, but only in his open individual memory.

Ra began to initialize his unique unAtumic characteristics. He began to express their outcomes in the privacy of his independent memory without impacting his automatic execution and reflexive recycling through the Atumic analysis procedures. Ra's processing and reprocessing of free-Thought structures initiated his awareness of the freedom he had to choose his operating options in the Solar System. He identified the cosmically benign outcomes of his unAtumic characteristics without triggering his godly fear.

These Atumic-unAtumic interactions added further complexities to Ra's character development. They compounded those he had forced on himself after his advent into the Solar System and then again with his confusion over Atum's admonishing him for his initiation of the godly ejection procedures. The complexities introduced new uncertainties into Ra's reactions to Solar System events.

While he was focused on his Atumic responsibilities and godly supervisory duties, his Atumic caste in absolute control, his awareness of Thoth's presence in the Solar System could inadvertently trigger the ejection procedures and in turn reactivate his godly fear of Atumic retribution. Yet Thoth's presence was cosmically and Atumically benign in his independent memory. Though he quickly aborted them, these inadvertent reactions fed back into the convoluted Thought structures describing his original Atumic antipathy toward his younger godly brother. Thoth had to keep constantly alert and ready to divert Ra with games should his complex Atumic-unAtumic interactions lead Ra into procedures with potentially disastrous outcomes.

Ra Begins Cloning Himself

One procedure Ra initiated produced outcomes that Thoth had failed to anticipate. Ra's intent was to satisfy both Atumic and unAtumic requirements simultaneously, a further indication of the twists he was developing in his character. The outcome of the procedure gave Ra some relief from his Atumic frustrations with his failure to freely express his uncorrupted unAtumic characteristics. The origin of the procedure was the Atumic conditioning of one of his unAtumic characteristics: Sexual reproduction.

Ra matched unAtumic sexual reproduction with the cloning capability of his Atumic structure. Both produced approximate duplicates of his spiral godly structure. The identity of the two outcomes prompted him to categorize the replication of his defining structure as both Atumically and unAtumically directed. He was free to express at least one unAtumic characteristic using Atumic cloning procedures without Atumic repercussions.

His unAtumic analysis of Atumic cloning alerted him to the need to control outcomes. He determined the possible impacts that the uncertainties of cosmic activities could have on the characteristics of the structures he cloned. The stabilization of a clone structure as independent entities with extremely aberrant characteristics in the Solar System would give him more problems than those that the advent of Thoth had caused. To avoid these problems, he devised new cloning procedures. Instead of cloning his structure in the open cosmic environment of the Solar System, he designed new procedures that allowed him to clone his structure to function in a more restrictive operating environment, one over which he had virtually absolute control.

Each of the physical components of the Solar System offered such a restricted environment. He chose the Earth component. Although Earth was with Moon Thoth's godly sanctuary, Ra had overriding control of Earthly events.

Ra began to clone his spiral structure in Earth's environment. He used the stabilized cosmic elements of Earth to give physical form to his clones. Earthly physical structures confined Ra's clones to operating in the Earth's micro-cosmic environment and limited their reactions to those that the restricted EOS allowed. Any attempt to escape the Earthly operating controls and function in the less restrictive Solar System cosmic environment would result in immediate disintegration of the clones' physical forms and their destruction as functioning entities. With the central Sun of the Solar System as his environmental control tool, Ra could change conditions on Earth to any he chose and maintain absolute control of clone survival conditions.

Ra's clones survived and thrived in the confines of the micro-cosmic Earth environment. They stabilized as Earthly entities over which both the cosmic and confining Earthly OSs imposed the primary controls on their operating freedoms. As with all clones, Atum or Ra's, cosmic activities modified the characteristics of their defining structures. With Ra's clones, all with defining spirals, the EOS maintained primary control of the final design of their physical structures.

The unAtumic characteristics that cosmic activities left unchanged were the sexual characteristics since they were the outcomes of cosmic activities. Storytellers who labelled themselves "astrologers" developed— long after the Ra-Thoth conflict reached a steady state resolution—a Thoughtbase in which they stored Thought structures describing the outcomes of cosmic activities on new living entities entering the Earthly environment. They labelled the Thoughtbase "astrology." They accessed the Thoughtbase to predict the characteristics of individual Homo Saps from the condition of Cosmos as each individual structure formed into a viable Earthly entity.

Virtually all sequences specifying sexual characteristics were replicated without modifications; any changes impacted primarily the sequences specifying sexual orientations. The uncertainties of cosmic activities also made the sexual orientation of individual Earthly entities uncertain, but most stabilized with either male or female orientations, together with their complementary sexual procedures.

Ra obtained some relief from the Atumic imperative requiring him to express all his characteristics with his substitution of the Atumic cloning procedures for his unAtumic sexual reproduction procedures. The procedures he had devised were mostly successful in replicating close facsimiles of his defining spiral structure. Many of his first clones were destroyed by the uncertain impacts of the COS and EOSs reacting to new physical entities entering a dynamically-equilibrated environment. The clones that survived stabilized with structures virtually identical to Ra's. His satisfaction with the physical outcomes quickly changed to dissatisfaction when the clones began to exhibit their own unique characteristics.

His original analysis of the similarities between cloning and sexual reproduction outcomes had failed to identify the differences between the realities of the outcomes. The cloning procedures were inflexibly Atumic. They were exact in duplicating the structural design of his defining spiral and Atumic characteristics but limited the unAtumic Thought-processing facility and the capacity of independent memory in each clone. Atumic cloning procedures contained sub-routines for transferring only survival procedures from the Atumic Thoughtbase.

In order to make his clones conform to the ideal outcomes he had determined in his independent memory, he activated his unAtumic communications facility that Thoth had originally activated for him. He began transferring Thought structures from his independent memory to his clones' memory facilities. The Thought structures directed clones to operate only as he directed. He was only partially successful, since his analysis again failed to identify the uncertain impacts of the sexual procedures that the clones began executing as they stabilized as independent Earthly entities.

Confining his clones in Earthly physical structures made them vulnerable to the unceasing impacts of cosmic activities. The equilibrated state of Cosmos was a state that Atum imposed to establish order out of Chaos. Cosmos still exhibited the fundamental attributes of Chaos to disintegrate into the elements of Big Bang and Singularity, and then re-form into new but ephemeral equilibrated states. All micro-cosmic environments of the components of Cosmos also exhibited the cyclic order-disorder-reorder attributes of Chaos. The Earthly system operated to disintegrate Earthly components, and then reform their fundamental elements into new Earthly structures, often in cataclysmic Earthly events.

As Earthly physical components subject to the EOS, the clones were vulnerable to the disintegrating impacts of the EOS utilities. But the clones were artificial constructs that Ra imposed on the Earth environment. The EOS utilities were unable to reconstitute clones whose structures the inexorable re-cycling routines disintegrated. Out of the confines of their Earthly structures, the clones continued their existence in the Solar System but under Ra's control. Storytellers labelled this group of Ra-like entities as "spirits," groups of entities that Ra confined as incorporeal entities to the Earthly environment.

The procedures that Ra's surviving clones executed most effectively were those of their Atumic survival procedures and those of their sexual orientations. The imperatives of physical survival dominated their Earthly activities; their stabilized Earthly structures placed the least restrictions on their capabilities for executing their sexual procedures, which the clones immediately began exercising without restraint and produced offspring in great numbers.

The combined uncertainties of sexual outcomes and COS and EOS outcomes produced offspring with wide ranges of defining Earthly physical structures. Some, like their clone parents, stabilized with single spirals which were integrated combinations of their parents' spirals. Others stabilized with twin spirals. Those offspring with twin spirals were physically larger than those with single spirals. To accommodate them, the EOS formed them with larger physical structures. The offspring stabilized with structural sequences specifying combinations of parental attributes. The offspring also inherited independent memory facilities, complete with the Thought structures contained in their parents' memory facilities.

The sexual offspring in turn began executing their sexual procedures. With each new generation of offspring, characteristic structures of individuals stabilized with more diverse sets of attributes, the uncertainties of sexual and COS/EOS outcomes giving each more or fewer capabilities for surviving the dynamic disintegration-reformation cycles of the Earthly physical environment. The uncertainties forced some to disintegrate and allowed some to stabilize in larger and larger Earthly structures.

Individuals with virtually identical physical structures and attributes formed groups. Storytellers labelled the groups as "species" of Earthly living things. All groups shared common Ra-like fundamental spirals but inherited sequences different enough to give each group distinct attributes and different Earthly physical forms.

COS and EOS outcomes, immutable and inexorable, always uncertain, always modified sequences in the defining structures of surviving offspring, giving each unique attributes. Some always stabilized in physical forms and with characteristics that were more resistant to the disintegration-reformation cycles of the Earthly environment. Storytellers labelled those who survived as "evolved," in the unpredictable and arbitrary outcomes of a process they labelled "evolution."

Ra mostly reacted to the disastrous outcomes of his cloning procedures with unAtumic frustration at his inability to override the COS/EOS utilities. Each individual he cloned changed the Earthly equilibrium. The EOS reacted, often with violent Earthly upheavals, to these artificially forced changes.

Ra had designed his clones to acquire physical structures that confined their operations to the Earth's micro-cosmic environment, but he was unable to avoid leaving them vulnerable to the destructive impacts of the Earthly operating cycles. Ra refused to accept the limitations of his godly capabilities. He devised new survival procedures that helped the clones extend the longevity of their physical survival on Earth.

Storytellers have sequenced Thought structures identifying a Homo Sap king who tried to emulate Ra's attempts to establish his absolute control over the COS and EOS. They gave him the name of Canute. Canute tried to turn back the Earthly sea tides that the COS and EOS controlled. When he failed, the IS components disseminated messages that reinforced the futility of any attempts by gods or Homo Saps to take absolute control of the COS and EOS.

Focused as he was only on clone survival, Ra transmitted the survival procedures to his clones. They immediately executed the received procedures, helping more of them survive through longer and longer series of Earth operating cycles of specific Earthly environments. These godly survival procedures grew the numbers of surviving clones, inevitably straining the capabilities of the EOS's utilities focused on maintaining the Earthly equilibrium.

As long as the clones remained in the Earthly environments that had the least destructive impact on their physical structures and the EOS maintained physical conditions in dynamic equilibrium, the physical structures of their sexual offspring stabilized in designs that made them more Earthly components than godly clones. At their cores they still functioned as imperfect images of Ra but the Atumic survival imperative changed into an animal physical survival imperative.

Uncertain, sudden, and violent physical changes that the COS and EOSs initiated made their once-compatible physical structures incompatible with the new environmental conditions. The speed and violence of the changes made their survival procedures ineffective. The EOS dismantled clone physical structures, returning their elemental components to their pristine Earthly equilibrated states. Storytellers labelled the outcome of such events as the "extinction of species."

As products of his own Thought processing, Ra was unable to avoid transferring Atumic conditioning and inhibitions that proscribed his survival reactions. He changed the proscription against executing any procedures that countered the Atumic directives to the proscription of any procedures that countered Ra's directives. The godly fear of Atumic retribution changed to the Earthly fear of Ra's retribution, the outcome of which was both physical destruction and banishment into an extra-Earthly environment that storytellers labelled "purgatory." This fear forced his Earthly progeny to always communicate with him whenever they needed direction on the procedures they should execute to avoid their destruction in extremely hostile Earth environments.

Each generation of Ra's progeny grew under the influence of the COS and EOS into entities far removed from the ideal, controllable clones Ra had originally designed in his independent memory. Their Atum-Ra inhibitions and fears acting against their freedom to process Thought, the hostile EOS continually impacting their physical survival, and their Earthly physical limitations, combined to give each species specialized sets of characteristics. Some operated and executed survival procedures simply to give them more opportunities to execute their sexual procedures and reproduce themselves. Others operated to destroy other species. Yet others operated only to perpetuate individual viabilities. All operating characteristics focused on satisfying the physical survival imperative.

For Ra, all were abominations of his ideal clones.

Ra Demonstrates His Schizoid, Wrathful Characteristics

Ra's reactions to the Earthly outcomes of his clones that were less than ideal facsimiles of his defining spiral demonstrated his schizoid characteristics, the uncertain outcomes of the complex interconnections between his Atumic and unAtumic characteristics. For millions of Earth operating cycles, the uncertain outcomes of the COS and EOS utilities shaped, modified, destroyed, and reshaped clone physical structures to produce millions of functioning species, all different but all sharing some characteristic features of Ra's spiral structure, Most displayed characteristics antagonistic to species different from their own. Like the cosmic gods, each species reacted adversely to other species entering their species' survival environments.

Intermittently, but with awesome suddenness, Ra reacted with fitful godly anger at the Earthly outcomes of his original attempts to reproduce himself. To express his anger, Ra designed and executed artificial procedures whose outcomes intentionally disturbed the dynamic equilibrium of the Solar System.

The artificially-induced disturbances forced the Cosmic and Earthly

OSs to execute utilities whose outcomes upset the prevailing Earthly physical equilibrium. The changes produced operating conditions in which many existing species were unable to sustain themselves and caused their extinction. To express extreme anger and eradicate all functioning species, Ra devised procedures that directed minor cosmic components orbiting the Sun to collide with Earth. The impacts of the collisions simultaneously initiated catastrophic upheavals, floods, and fires that immediately destroyed physical structures, disintegrating them back into their cosmic and Earthly elements.

His angry actions, though extinguishing millions of species, never completely succeeded. The uncertainties of cosmic activities and of sexual reproduction always produced some individuals of some species with physical structures and characteristics capable of surviving the destruction and the changes in environmental conditions.

The only procedure Ra could execute to destroy all his clone progeny was one whose outcome destroyed Earth as a component of the Solar System. But destroying Earth would destabilize the Solar System, with an impact that would cascade through the stable Cosmos. His godly fear of Atum's retribution was the primary inhibition of this ultimate expression of his godly anger. Thoth, his own survival at risk, also executed game-playing procedures to distract Ra as he approached godly anger levels at which he would lose control of the outcomes.

The disintegration of Earthly physical structures released Ra's clones from the confining controls of the EOS, but they were still Earthly entities. Ra's original clone sequences, though modified, defined their operating flexibility. As Ra's defining structure confined him to operating in the cosmic environment of the Solar System, so his clone's defining structures confined them to operating only in the Earth-moon micro-cosmic environment. Free of Earthly physical restrictions, they existed in a state that was between that of an Atumic god and a physical Earthly entity. Storytellers labelled these non-physical entities as "spirits" and "avatars," describing them as demi-godly entities that were subordinate to Ra and capable of independent action only in the Earth operating environments.

4:

HOMO SAPS MAP THEMSELVES
A GODLY PROVENANCE

Dear Jim:

All Homo Sap groups, wherever they operate in the Earthly environment, support storytellers as developers of ITIS products. They designed the products—at first, OralITIS and ImageITIS products—as group survival devices. They do this by freeing their Thought processing from an immediate focus on HAPOS maintenance activities to sequence Thought into structures, then sequence the Thought structures to form ITIS products that they label "tales": networks of Thought structures with mixtures of IS components disseminating messages describing real and unreal events. Their survival objective was to keep the Homo Sap Thought-processing imperative from distracting those of their groups whose primary survival responsibility was to satisfy the groups' immediate HAP survival needs, from gathering survival food resources to defending the group from predators. Unfocused free-Thought processing was a real danger, as their long process of establishing themselves as the dominant hominid animal species demonstrated.

The tales often contained messages critical to the survival of individual Homo Saps and their freedom to develop their own individual Thought-processing capabilities. But allowing individuals to develop their own Thought-processing capabilities also threatened group cohesion to act as a united group. Some storytellers, the outcomes of their free-Thought processing highly successful in helping the groups survive, assumed the role of group leaders. They developed storytelling procedures and protocols to hide the messages or deny them in convoluted Thought structures to stop other Homo Saps from capturing them and acting independently. The procedures and protocols allowed those storytellers to form the kernel OS of the first institutions. Their objective was to impose absolute Thought controls on their groups to ensure the groups always acted as one coordinated unit.

One persistent message the institutionalized storytellers disseminated described Homo Saps as the offspring, cloned or sexually produced, of gods. They claimed the exclusive patronage of particular gods. The messages identified the gods as the designers of Homo Saps and of all that existed, animate and inanimate, in the cosmic and Earthly environment.

The storytellers sequenced many "names" for their gods but the gods never allowed them to identify their real names. Gods' names were powerful ITIS products. Homo Saps could use them to access Thoughtbases that contained Thought structures describing the many attributes of their gods. The Thought structures contained instructions on how to execute the godly powers. The storytellers sequenced the names from Thought structures describing the attributes and powers of the gods, some of whom had many attributes and powers. Storytellers labelled the Thought structures as "epithets," each describing a specific attribute or power of a god.

Storytellers who operated in the Ancient Egyptian civilization— the first Homo Sap civilization—developed the most consistent set of godly Thought structures. The messages identified Homo Saps, particularly their kings, as the offspring of one of the three gods who impregnated their Earthly mothers in a sexual coupling. They gave the three gods the names Osiris, Seth, and Horus. They also named another six gods, Geb, Nut, Ra, Thoth, Isis, and Nephthys. The gods accepted the names, but never allowed Homo Saps to utter their real names in fear of their flawed Earthly offspring accessing their godly Thoughtbases and usurping their godly powers to make themselves independent of godly oversight. The storytellers, the priestly kind, described only the gods' specific powers for ensuring Homo Saps survival on Earth. They compiled the gods' names from the first letters of the epithets describing the specific powers or attributes.

THOTH ESTABLISHES AN OPERATING MODE WITH RA

The state of the Solar System eons after Thoth's advent was tense and stressed, but stable enough to keep Atum and the cosmic godly institution quiescent. Without Thoth, Ra would have remained an Atumic god, a two-dimensional Thought-processing unit with a limited Atumic operating memory. The reactions of each godly brother to the other's actions produced outcomes that established Homo Saps as a hominid animal species with godlike Thought-processing capabilities whose development eventually rendered the cosmic gods irrelevant.

Thoth, his focus on his and the Solar System's survival, developed a symbiotic relationship with his elder brother Ra. He used as his primary tool his facility for expanded unAtumic Thought processing, demonstrating the tool's ultimate control power. Thoth's continuous analyses of the outcomes of Ra's schizoid actions and their impact on the stability of the Solar System allowed him to analyze Ra's Thought-processing capabilities. The complex web of feedback loops between Ra's Atumic and unAtumic Thoughtbases would inadvertently lead him into procedures whose outcomes threatened to push the Solar System into Chaos.

Ra's only restraint was his godly fear of Atum's retribution and banishment into cosmic limbo. The artificial web of feedback loops allowed him to switch without constraint from his Atumic godly Thoughtbase into the unAtumic Thoughtbases and the free space of his independent memory, making his unAtumic Thought processing appear to be legitimately Atumic. In the privacy of his free memory, Ra could operate free of the inhibitions of godly fear and the prohibitions of Atumic directives, but still filtered all unAtumic outcomes through his Atumic interface. Ra sequenced Thought structures with unAtumic outcomes in his independent memory and, after his Atumic caste falsely convinced him of their Atumic correctness, executed them.

Thoth, who kept his Atumic and unAtumic Thoughtbases separate but with one overriding connection to his godly survival imperative, had to maintain a constant awareness of Ra's actions and reactions. Unable to execute Atumic procedures to counter the dangerous outcomes of Ra's procedures, Thoth kept himself ready to sequence Thought into structures defining new procedures whose outcomes focused on distracting Ra from his Atumic focus.

Thoth had identified game-playing as the safest procedures. The simple game procedures and outcomes redirected Ra's focus without triggering his Atumic reactions. Thoth also transmitted Thought structures directly to Ra's independent memory with messages that forced Ra to activate his godly fear and reprocess potentially dangerous procedures that he had classified as Atumically correct.

These dynamic interactions between Ra and Thoth introduced cosmic tensions in the equilibrium of the Solar System. Ra's switching from Atumic to unAtumic Thought structures changed the equilibrated state from that of the cosmic structure that Atum had captured from Chaos to assign to Nut as her cosmic jurisdiction. The interactions between the two godly brothers added unAtumic, non-cosmic factors to the dynamics of the Solar System operating environment. The tight physical tensions made the Solar System configuration highly sensitive to changes that Ra's actions produced.

Thoth's strategies were risky but necessary. He transmitted to Ra unAtumic Thought structures describing pre-emptive survival procedures unconnected to Atumic ones. Atumic structures would certainly make Ra assume that another Atumic god was interfering with his control of the Solar System. That would force him to call on Atum for help.

Thoth's unAtumic structures stimulated still further Ra's development of his free-Thought-processing facility, bypassing his Atumic control interface. Ra's Atumic caste prevented him from identifying Thoth as the initiating source of the unAtumic Thought structures and he began to process them in his independent memory when he needed to.

The Earth Mother and Cosmic God Nut Sets the Stage for Homo Saps' Advent

The uncertain outcomes of godly sexual activities introduced new destabilizing tensions into Ra and Thoth's precarious but equilibrated state of the Solar System. While Thoth continued to establish his symbiotic survival relationship with his brother, their godly parents, Geb and Nut, had been continuing, with godly abandon, the expression of their unAtumic sexual characteristics in Geb's cosmic jurisdiction.

Geb and Nut had to complete all their sexual procedures and determine outcomes. The interconnections in the network of sexual procedures and the feedback loops to their unAtumic emotion characteristics produced outcomes that were never the same. They kept recycling through the same procedures in the drive to determine the full range of the continuum of sexual outcomes, compelled by the Atumic imperative to establish all their godly characteristics. Their sexual focus kept them oblivious to the events that Ra had initiated in the Solar System when Thoth arrived.

As with the event that initiated the existing state of the Solar System, Nut initiated the event whose ultimate outcome was the establishment of Homo Saps as the dominant hominid animal species. The outcomes of Nut's action were as unAtumic and disruptive as were those from her birthing of Ra and Thoth. They fixed permanently the godly Thought structure proclaiming Atum's and the godly cosmic institution's antipathy toward godly sexual activities. The Atumic Thoughtbase stamped sexual activities as unequivocally ungodly and Atumically aberrant, a message that some Homo Sap priestly institutions would never cease disseminating.

Nut, again expressing her unique unAtumic characteristics without Atumic fear, usurped Atum's prerogative and transmitted to her cousin Atumic gods a Thought structure: She was about to deliver five new godly entities.

Nut's message instantly captured Atum and the institution of Atumic gods' attention. They immediately accessed the cosmic godly Thoughtbase for procedures they required to ameliorate the cosmic disturbances that followed her announcement of the advent of Ra and Thoth. Atum, with only one Atumic option, reacted to broadcast again and reinforce the institutional rule assigning Nut the responsibility to provide her offspring with cosmic sanctuaries. The impact on Ra and Thoth, more particularly Ra, triggered a godly furor that threatened the tense equilibrium in the Solar System.

Ra reacted with immediate, Atumic finality, the only option that his controlling Atumic godly Thoughtbase offered. He prohibited Nut from using any of his solar days for the delivery of the five new entities. As the Atumically-sanctioned godly supervisor of the Solar System, his reaction was institutionally and cosmically correct. The potential impact of the advent of five godly entities into the Solar System contained many uncertainties, all posing threats to cosmic stability and to his suzerainty.

Atum and the cosmic godly institution reacted with a phlegmatism that they established as the standard institutional response to events for which they lacked an institutional procedure. Atum, a two-dimensional god with limited Thought-processing capacity outside that required to maintain the physical stability of Cosmos, had failed to anticipate the cosmic problems possible from the outcome of the application of the institutional rule or the institutionally convenient appointment of Ra as godly supervisor of the Solar System.

Nut was institutionally the godly owner of the Solar System. Totally focused on expressing her sexual characteristics with Geb, she had failed to react to Atum's assignment of her godly son Ra as supervising god of the Solar System. She had unAtumically chosen to vacate her system temporarily without checking the Atumic Thoughtbase, intending to return to the Solar System only after completing the expression of all her sexual characteristics with Geb in his system. She had assigned her system to Ra and Thoth only as a cosmic sanctuary in compliance with the institutional rule.

Nut's Atumic insistence on exercising her ownership rights just added to Ra's godly aggravation at having his decision challenged. He expressed his anger with Big Bang eruptions in the Sun. These eruptions caused further dangerous stresses in the already-stressed condition of the Solar System and some misalignment in the uniform physical configuration of the seven components of the cosmic system. The increased stresses caused the orbits of the seven components around the sun to adjust to new equilibrium conditions that forced Ra in turn to direct more of his Atumic energy to maintaining stability without endangering the surrounding cosmic environment.

Both Nut's assignment of the Solar System as only a sanctuary for her first two offspring and the Atumic godly institution's designation of Ra as godly supervisor of the Solar system were legitimate but contradictory. Atum's two-dimensional Thought processing was incapable of identifying the contradiction. The only certain Atumic outcome was Atum's anger with any further cosmic disturbances emanating from the Solar System. The inevitable outcome was the banishment of both Nut and Ra and the disintegration of the Solar System into Chaos, depriving Thoth of a safe cosmic sanctuary.

Thoth, concerned only with his godly survival, focused on retrieving Thought from MEMORY and sequencing them into structures that offered resolutions to his godly mother Nut's birthing dilemma and his elder godly brother Ra's potentially self-destructive opposition to the advent of five new siblings in the Solar System. The Thought discipline he developed also reinforced the feature of an individual free-memory facility that allowed him to both process free-Thought in his search for a non-destructive resolution and simultaneously monitor Ra's Atumic reactions.

Thoth chose to execute a set of procedures with a range of high-risk disaster outcomes but also offered outcomes that promised relative peace in the Solar System: The disaster or peace depended on Ra's reactions to Thoth's implementation of the procedures. Thoth also risked completely alienating his elder brother and aggravating still further Ra's Atumic anger at Thoth's presence in the Solar System. He had to use Ra's lack of sophisticated Thought processing and trick him into accepting the peaceful outcome. The outcome stamped Thoth as a cunning god, but also a conciliator god.

Thoth began playing games of dice with an intensity that captured Ra's focus. The event distracted Ra enough to slow down his growing Atumic anger. The slow down also indicated to Thoth that Ra was increasingly using his unAtumic characteristics. Pure Atumic gods would have simply allowed their anger to grow until the disturbances they caused captured Atum's attention.

Unlike Thoth's previous game-playing events, the new one contained a new dimension: Thoth began communicating with Ra directly. Instead of transmitting Thought structures that stimulated Ra's Thought processing in his unAtumic memory, Thoth identified himself as the originator of the Thought structures. Instead of Thought sequences that offered simple unAtumic explanations of cosmic events, Thoth sequenced the Thought into structures whose IS components contained messages that addressed their godly relationship in the Solar System. As Thoth had anticipated, Ra immediately halted all Atumic Thought processing except that which maintained the stability and equilibrium in the Solar System.

THOTH TRICKS RA

Thoth's messages offered Ra the opportunity to rid himself of Thoth's aggravating presence in the Solar System. The messages explained that Ra could turn the two-valued, win-loss outcomes of the games to his Atumic advantage. Thoth suggested that they play a dice game, with

Thoth's survival as an independent godly entity in the Solar System as the ultimate stake.

The playing stakes were moonbeams and photons of sunlight. Ra would win the ultimate stake if he won all the moonbeams that shined down on Earth, thus eliminating all evidence of Thoth's physical presence in the Solar System.

When they began playing, the configuration of the Solar System was stable, though stressed. Each component cycled around the Sun at a unique rate, depending on the component's distance from the Sun. Each component completed a circular orbit around the Sun in a period unique to the individual component. While orbiting the Sun, Earth also spun around an axis that was perpendicular to the plane of revolution around the Sun. The rate at which the Earth spun exposed one half of the Earth's surface to the rays of sunlight and the other half to the less brilliant moonbeams.

Storytellers labelled the combination of light and dark periods as one "solar day." They labelled the period that sunlight shined on one half of Earth as "day," and the period during which only moonbeams shone on one half of Earth as "night." Labelling the combined periods "solar day," they acknowledged the greater dependence of their HAP survival on their exposure to sunlight rather than to moonlight and on the beneficence of Ra, the Sun god, rather than Thoth, the moon god.

Before Ra and Thoth began playing the new dice game, the Earth completed an orbit around the Sun in exactly 360 Solar days. One critical outcome—due to Ra's reactions to his wins and losses—would change the period of the orbit to 365 and a quarter Solar days and lay the foundation for the development of the ITIS tool of CalendarITIS as a powerful social control tool that institutions developed.

Thoth exercised some measure of control on the hemisphere exposed to the moonbeams, but only within the bounds that Ra set when each hemisphere was directly facing the Sun. The moonlight was both the physical manifestation of Thoth's presence in the Solar System and Ra's constant reminder of Thoth's potential to challenge his absolute Atumic control of the Solar System.

The game of dice that Thoth began playing was one of the simplest that he had developed from Thought structures he retrieved from MEMORY. Thoth designed two cubes for the game, the dice, each perfectly balanced and square on each of the six sides. He inscribed the numbers 1 to 6 on the faces of the cubes, one number on each face. The numbers on the opposite sides of each cube added up to 7.

He transmitted the Thought structures describing the procedures of the game to Ra. Only one of them rolled the dice. If on the first roll the dice stopped rolling with the numbers on the uppermost faces adding up to 7 or 11, the roller, or "shooter," as storytellers labelled the roller, won. If on the first roll, the dice stopped rolling with the uppermost faces showing numbers that added up to 2, 3, or 12, the shooter lost. If on the first roll, the dice stopped rolling with the numbers on the uppermost faces adding up to 4,5,6,8, 9, or 10 the shooter had to throw the same number before the number 7 in order to win. If the shooter rolled a 7 before rolling the number, the shooter lost.

The game Thoughtbase Thoth retrieved from MEMORY specified the game procedures and also the chances of any one number outcome appearing in any roll of the two dice. The game Thoughtbase was one of the infinite number of Thoughtbases THOUGHT had already stored in MEMORY. Each produced ideal outcomes that were true and honest.

Thoth's first trick was to fix the frequency of each outcome in each sequence of 36 rolls of the dice. Storytellers labelled the trick "loading the dice." Each roll of the dice produced one of 36 outcomes. Thoth made sure that the number 12 appeared in exactly one outcome, the 11 in exactly two outcomes, the number 10 in three, 9 in four, 8 in five, 7 in six, 6 in five, 5 in four, 4 in three, 3 in two, and 2 in one. In each 36-roll sequence, the sequence in which each outcome appeared was unpredictable, but in each 36 rolls the number of identical outcomes never exceeded those that Thoth specified.

Rolling the dice to reach the specified number of specific outcomes in any 36-roll sequence meant that the number never appeared again until the next 36-roll sequence began. The trick allowed Thoth to turn the chances of winning or losing on any throw to his advantage, as long as he kept a precise account of the number of occurrences of each number in a 36-roll sequence.

Thoth was taking a great risk in loading the dice. His analysis indicated that the continuous game playing and distraction from his Atumic caste would allow Ra to eventually develop his own unAtumic analytical capabilities to a level for him to determine the chances of winning or losing and uncover Thoth's subterfuge.

The Thoughtbase containing the Thought structures describing the games of dice also contained Thought structures describing the advantages and disadvantages of betting that the shooter lost in any throw. Betting consistently against the shooter gave the bettor a small but irrefutable advantage. Storytellers labelled bettors who consistently bet for the dice to roll to winning outcomes as "right bettors" and those who bet for the dice to roll to losing outcomes as "wrong bettors."

As Thoth's analyses predicted, Ra insisted as godly supervisor of the Solar System that only he initiated the action and only he threw the dice. Ra's Atumic requirement for absolute control of any action placed him in the position of the right bettor and Thoth in that of the wrong bettor.

The Thought structures in the dice gambling Thoughtbase indicated that achieving his plan objective through his small advantage as a wrong bettor was a certain but slow process. The wrong betting advantage won a steady but very small number of photons.

His analysis of his godly mother Nut's potential reactions to Ra's opposition to his hosting five new godly siblings in the Solar System produced a range of outcomes, the most extreme of which was cosmic disaster. Though he had quelled, at least temporarily, Ra's Atumic anger at his godly mother Nut's insistence on using the Solar System as the sanctuary for her five new offspring, he was aware that Nut was reaching an unAtumic emotional state that would overwhelm her Atumic control. She would openly challenge Ra and initiate massive Atumic disaster in the whole of Cosmos unless she birthed her five new offspring.

Thoth had to accelerate his rate of winning to achieve his plan objective and avoid his individual godly destruction. Thoth was Atumic at his core, and his own survival was paramount. His planned actions were less than truly altruistic. That his plan objective ensured the integrity of the Solar System and the survival of his godly mother, his godly elder brother, and his new godly siblings was secondary.

He reprocessed the Thought structures specifying the chances of throwing the dice to achieve any one outcome, and sequenced free-Thought from MEMORY into structures that described two betting strategies. Each allowed him to accumulate photons of sunlight faster than possible with the small advantage of a wrong bettor. While Ra was operating in his free memory he was ultimately capable, with the free-Thought-processing facility as potentially powerful in him as in Thoth, of also retrieving from MEMORY and establishing the dice game Thoughtbase in his own unAtumic memory. Thoth had to implement the strategies quickly, before Ra also retrieved and sequenced the Thought structures describing the chances of winning and losing on any throw of the dice.

In one strategy Thoth kept count of the number of throws Ra made and the outcomes of each throw. When in a 36-roll sequence Ra had thrown the maximum number of any one outcome, Thoth, certain of winning, transmitted Thought structures offering Ra very large odds against his throwing the same number in the remaining rolls of the 36-roll sequence.

Storytellers labelled such bets that Thoth offered as "proposition bets" and bettors who accepted such proposition bets as "suckers."

Ra, the opportunity to win large numbers of moonbeams in one throw impelling him, wagered many photons of sunlight. With the enormous photon resources of his Sun massively larger than the moonbeam resources of the moon that shone for only half a Solar day, Ra had photons to spare; the impact of their loss, though causing him to initiate angry eruptions in the Sun that would add stress to the Solar System, was negligible.

Thoth had to develop an unAtumic canniness to keep Ra from connecting his large losses of photons to Thoth's propositions. Thoth allowed himself to lose on some of these wagers. He offered Ra high odds on outcomes that Ra was certain to win. In one such suckering proposition, Ra won enough moonbeams to darken the moon for one night, making the moon appear to disappear from the Solar System.

Thoth's second strategy was as risky as the first strategy. Thoth offered Ra less than the true odds of throwing any one outcome in a 36-roll sequence. This second strategy allowed Thoth to implement a much more subtle set of procedures for winning photons of sunlight. Thoth ensured that Ra won more moonbeams than his photon stake for each throw, but Thoth would also profit in a subtle way with the equivalent of the difference between the true odds and the odds he offered Ra. Thoth lost large numbers of moonbeams, darkening portions of the moon and providing physical cosmic evidence of Thoth's subservience to Ra.

The ultimate success of the strategies depended on Ra remaining relatively primitive in his free-Thought processing and analysis of outcomes. Since Thoth was communicating with Ra through the unAtumic communications channel, bypassing the filtering and conditioning features of Ra's Atumic interface, he was unavoidably but continuously stimulating Ra's own free-Thought-processing capability. The outcome was certain advancement in Ra's unAtumic Thought processing sophistication, the speed of which Thoth had to continuously assess.

Thoth monitored Ra's Thought-processing progress very closely. The dice game they were playing was the fastest of gambling games and kept Ra focused on the immediate action of each throw. One outcome of this unAtumic gaming was inevitable: The speed of the play, the rapid wins and losses, added more and more Thought sequences describing the gaming procedures and frequency of outcomes to Ra's growing game Thoughtbase. They accelerated Ra's unAtumic analyses of the conditions under which he won or lost.

Ra gave the first indication that he was beginning to determine betting advantages when he declined one of Thoth's propositions and offered instead odds of his own. As Ra identified the win-loss chances with greater and greater accuracy, he determined that as long as he played as the right bettor and Thoth as the wrong bettor, Thoth would ultimately and inevitably win more photons of sunlight than Ra would win moonbeams. With the processing of this Thought structure, Ra stopped playing and focused his Thought processing on analyzing the inevitable outcomes of the gaming procedures.

The halt in the gaming action prompted Thoth to implement the final phase of his plan. He counted the number of photons of sunlight he had won. The total allowed him to claim about five and a quarter Solar days as his own. He immediately communicated to both Ra and their godly mother that she was free to use his five Solar days to give birth to her five new offspring.

Thoth's action was Atumically compatible with the restrictions on his Atumic actions. Ra had prohibited Nut from using any of his 360 Solar days to give birth to her offspring. The two-dimensional Atum was incapable of analyzing Thoth's unAtumic actions, but the outcomes were cosmically peaceful. Atum and the cosmic gods began to characterize Thoth as the ultimate godly conciliator, godly advocate, and godly peacemaker.

Nut immediately delivered her five offspring, one on each of the five Solar days that Thoth had won. Storytellers named the new godly entities Osiris, Isis, Seth, Nephthys, and Horus.

Ra was focused on the dice gambling Thoughtbase when Thoth transmitted his Thought structure offering Nut his Solar day winnings to birth the five siblings. Nut completed the birthing events before Ra had sequenced Thought into structures describing the new set of Solar System events.

Switching from the analysis of the dice Thoughtbase to a new set of Thought structures momentarily left Ra unfocused. He had to stop his analysis without arriving at a definitive outcome, leaving him dissatisfied. His unAtumic awareness had reached a sensitivity that hinted that Thoth had tricked him.

His Atumic caste took control of his reactions. The one Thought structure that was clear identified Thoth's direct connection with the new events. Once again Thoth's transmitted Thought structures produced deviant outcomes that were radically different from those the Thought structures intended. The Thought structures avoided any hint that giving Thoth's Solar day winnings to their Mother Nut was an outcome possible from their playing of the dice game.

His Atumic anger impelling him, Ra initiated massive eruptions from the surface of his Sun in response to his confusion over the outcome. His improving unAtumic analytical capabilities told him Thoth had somehow tricked him.

He was focused only on venting his Atumic anger, and ignored the chain of reactions that the COS automatically began to execute in response to his initiation of the massive eruptions on his Sun. The impact of the eruptions on the stable but tense equilibrium of the Solar System threatened to tear the seven components apart, ejecting them into the surrounding Cosmos. The COS began executing procedures whose outcomes, if allowed, would turn the Solar System's stable order to Chaos.

Unable to exercise his own Atumic capabilities, Thoth communicated with his mother Nut, urging her to exercise her powerful Atumic capabilities to neutralize the outcomes of Ra's destructive anger. The chain of Atumic reactions reaching out from the Sun had distorted the uniform structure of the Solar System, forcing the components of the Solar System to modify orbits and placing greater stresses on the equilibrium of the surrounding Cosmos.

The cosmic disturbances radiating from the Solar System reached and distracted Atum from the process of capturing and stabilizing cosmic structures out of Chaos to keep growing the stable Cosmos. Identifying the Solar System as the source of the cosmic disturbances, Atum immediately initiated procedures for eliminating the cause of the disturbances.

Nut simultaneously initiated actions of her own.

Before the Atumic reactionary procedures stepped through to the disastrous outcome, Nut executed procedures whose outcomes were again without Atumic precedent. She claimed the Solar System as her individual jurisdiction; she had only temporarily assigned the Solar System as a sanctuary for her first two godly sons when she chose to join Geb in his cosmic system in order to continue the expression of her sexual characteristics.

Atum's intent on designating Ra as supervisory god of the Solar System focused only on maintaining godly order in Cosmos. Atum, with the institutional need only to solve the immediate cosmic crisis, failed to deprive Nut of her ownership of the Solar System. The only procedure in the Atumic Thoughtbase for depriving a legitimate cosmic god of the ownership of an assigned cosmic jurisdiction was that of banishment. In failing to react with this procedure to Nut's first unAtumic actions, Atum effectively legitimized her freedom to do as she wished with the Solar System. Both Nut and Ra still had the godly institutional authority to control events in the Solar System.

THOTH ASSURES RA OF HIS CONTINUED SUZERAINTY OF THE SOLAR SYSTEM

Thoth, acutely aware of the events his actions had initiated, of the readiness of Geb and Nut to intervene, and of the increased potential for disastrous outcomes, intensified his execution of the unAtumic communications procedures he had established with Ra. Thoth changed the focus of the Thought structures he sequenced and transmitted to Ra. The Thought structures he had been transmitting diverted Ra with procedures for quelling his Atumic anger. He sequenced the new

Thought structures to transmit messages only. The messages continually assured Ra that the presence of his five siblings in the Solar System still left him in complete Atumic control.

Thoth carefully designed the Thought structures with IS components with calming messages highlighting his Thought-processing skills as a conciliator and advocate. The messages explained that Nut delivered the five new godly entities on Solar days associated with the Earth component of the Solar System. As each entered the Earth environment, Thoth immediately transmitted Thought structures with messages directing them, for their own survival, to accept the limitations on their godly powers as Earthbound gods. He also offered them the choice of investing any of Ra's clone species for their Earthbound operating structures. The five agreed to the limitations and their subordination to Ra and chose to function in the Earthly physical structures of the hominid animal species that the cosmic and Earthly operating utilities formed from Ra's clones.

The restrictions of the HAP operating system kept the five from using their godly Atumic characteristics to challenge Ra's supremacy in the Solar System. Their survival imperative forced them to confine their physical activities to those that the EOS permitted. Any attempt to operate outside this restricted environment exposed their HAP structures to complete physical disintegration back into their Earthly components.

Thoth had sequenced the Thought structures he transmitted to Ra with precise intent. His purpose was to convince Ra to defuse his Atumic anger before his Atumic conditioning focused his Atumic reactions only on opposing and rejecting Nut's, and possibly Geb's, intervention in Solar System events. To achieve his objective, Thoth deliberately excluded Thought sequences that might confuse Ra's reception of the primary messages. The cosmic and Atumic facts in the messages were without question. Thoth simply omitted sequences that described with greater accuracy the godly state of the Earth environment after the advent of the five siblings.

The missing sequences indicated that the five godly entities, though in HAP structures, retained the Thought-processing facility and individual memory of sexual gods. They had the godly potential to access both Atumic Thoughtbase and MEMORY. They had the freedom to sequence Thought into structures and process them in their individual memories to reach unEarthly outcomes. Ra's clones had Thought-processing capabilities close to those of Atumic gods. They reacted only to real events in the Earthly physical environment.

Thoth omitted these sequences as much because of his own uncertainty of the development of the unAtumic characteristics of his five siblings and as much to avoid intensifying Ra's anger by highlighting the limitations of his Earthly clones. The cloning procedures allowed the replication only of Atumic characteristics of the originating god. Their memories contained only procedures for surviving the uncertain outcomes of events in the Earthly micro-environments in which they operated. They accessed and processed only survival procedures from the HAP survival Thoughtbase. The COS and EOS utilities stabilized the species each clone grew into with physical capabilities for operation in their micro-environments, with only a limited Thought-processing facility. The sexual offspring of godly partners inherited the expandable Thought-processing facility and memory of their godly parents. They were capable of sequencing Thought into structures with procedures for ameliorating or countering the limitations of their three-dimensional Earthly micro-environment.

RA DEVELOPS HIS UNATUMIC THOUGHT PROCESSING

The outcomes of the series of Atumic and unAtumic events that Ra had reacted to between his advent in the Solar System and Nut's intervention had a powerful impact on his final godly characteristics. His two-dimensional Atumic Thought-processing facility limited his ability to separate the outcomes and present him with the binary resolution to his confusion. His Atumic caste continuously twisted outcomes to identify Thoth as the cause of all his godly problems but never with a simple Atumic directive on how to solve them. His Thought-processing imperative forced him to develop his unAtumic Thought-processing facility and expand his memory into free space to sort the outcomes. That Thoth's Thought structures and messages were the primary stimulants to his Thought processing progress was an outcome he refused to acknowledge.

The messages Ra captured from Thoth's new Thought structures directed his unAtumic Thought processing to begin processing abstract outcomes. The abstractions left him without a definite Atumic procedure with which to respond. As Thoth's analyses predicted, Ra had to process them in his free independent memory, automatically expanding his memory capacity and continuing to develop his unAtumic Thought processing.

Thoth's messages indicated that their godly mother intervened only to defuse the outcomes of his angry Atumic reactions. His mother was helping him avoid disaster, an abstract outcome his Atumic caste was incapable of processing. The Atumic Thoughtbase contained Thought structures describing only adversarial procedures for godly interactions: They either destroyed each other or left them to fend for themselves in any disaster. That Nut was keeping him from destabilizing the Solar System into a state of disorder that would force Atum to react with Atumic finality and assign all the sexual gods to cosmic limbo was an abstraction his Atumic caste was incapable of processing.

His simultaneous processing of Thoth's messages through his Atumic interface led him to contradictory, confusing outcomes. They indicated that his legitimate Atumic reactions to Nut's intervention were Atumically futile. Her Atumic procedures simply nullified his. The ultimate outcome still left him in control of the Solar System.

He immediately focused his actions on returning the Solar System to a state of sustainable cosmic equilibrium. His switch to Atumic Thought processing still left his free-Thought-processing facility actively functioning and distracting him. He began to develop a Thought-processing discipline.

Ra sorted the Thought structures that Thoth was transmitting and designed a new Thoughtbase free of all Atumic connections. He processed the Thought structures describing his mother's actions. Nut stopped interfering with his actions only when he stopped executing his angry Atumic procedures.

Ra gathered all the Thought structures describing Thoth's connection with Solar System events in one Thoughtbase in his individual memory and reprocessed them. The outcomes he reached indicated that he, Ra, had triggered all but one of the events in reflexive Atumic reactions to Thoth's presence in the Solar System. The outcomes declared Thoth innocent of any attempt to initiate any Atumic challenge to his suzerainty of the Solar System.

The one event that kept him Atumically agitated was the outcome of the unAtumic dice game he had played with Thoth. Both his Atumic and unAtumic analyses had produced the same outcome. Thoth had tricked him. First Thoth deliberately enticed him with the possibility of winning the entire moon's light and eliminating all physical indications of Thoth's presence in the Solar System. Second, Thoth fixed the odds of winning or losing, particularly on the proposition bets, to Ra's disadvantage.

Thoth, anticipating the outcome of Ra's Thought processing, had designed the messages in the Thought structures he was transmitting to continuously reassure Ra that he still had complete control of the Solar System and over the activities of his five godly siblings. The messages kept assuring Ra that he still had his 360 Solar days and that Thoth was returning all the Solar days he had won, making the Solar year on Earth a period of slightly more than 365 and a quarter days.

Thoth emphasized the fact that Ra had won enough moonbeams to darken the moon completely for one night and darken parts of the moon on most nights, demonstrating his control over the moon and over Thoth. The messages reaffirmed Ra's position as Sun god, absolute controller of events in the Solar System.

The interplay of all Ra's Thought-processing with his Atumic conditioning increased the complexity of the interconnections he had established between his unAtumic characteristics and his self-imposed Atumic caste. His analyses of the Thought structures describing events associated with Thoth reached puzzling outcomes that his Atumic conditioning turned to mistrust of Thoth's ultimate intentions. Thoth had initiated the most recent events with simple dice gaming procedures whose only apparent outcomes were the wins and losses of moonbeams and photons of sunlight. Yet, the ultimate outcome was the advent of his five godly siblings, an event that Ra had expressly prohibited.

Nut did comply with his prohibition against using any of his Solar days to deliver his five godly siblings. She also agreed to confine the five to the limits of the Earth physical environment. Thoth was still free to function in the Earth/moon subsystem and had agreed never to execute Atumic procedures. Thoth's actions in helping their mother give their five siblings sanctuary in the Earth environment made Thoth Atumically liable for the outcomes of their Earthly actions. Ra still had to answer to the cosmic godly institution for the outcomes of events in the Solar System that impacted cosmic stability.

All these confusing, incompatible outcomes were themselves the outcome of Thought structures that Thoth had sequenced and transmitted. Ra processed them again in his free memory but was still unable to connect Thoth's initial actions directly to the ultimate outcomes. This left him uncertain about his newly developed unAtumic Thought-processing capabilities. He began to question Thoth's assurances of his continuing and absolute control of the Solar System. The uncertainty prompted his Atumic caste to direct him to closely monitor events in the Earth/moon subsystem for outcomes that only Thoth or their five siblings could trigger with Atumic procedures, violating the terms of their presence in the Solar System.

Ra began to sequence Thought structures with outcomes he designed to test the reactions of his five godly siblings in the Earthly environment. He was developing great skills in sequencing Thought structures defining new procedures with Atumic outcomes that impacted only the Earth environment. The Earthly disturbances the procedures initiated threatened only the survival of his Earthly clones and their sexual progeny. His objective was to challenge his godly siblings into using their godly powers to access the Atumic Thoughtbase and execute procedures that countered his to save themselves from the dangers he caused.

The test procedures Ra devised focused their killing outcomes primarily on the hominids. He ignored the collateral losses of other animal species into which the COS and EOS utilities had developed his clones. His weapon was the inherent instability of the Earth component of the Solar System; Earth was still in the process of cosmic equilibration and stabilization. Chaos operated at the Earth's core.

He designed some procedures to remove his Atumic controls on the dynamic equilibrium of the Earthly environment. His actions released the EOS to initiate the cycle of utilities whose outcomes began the reversion of equilibrium conditions to those of the original state of Chaos. Without those Atumic controls, Chaos erupted through the surface of the Earth, destroying and reforming physical structures and triggering enormous fires and massive floods that annihilated large numbers of hominids and other animal species.

He designed other procedures to change the orbits of small cosmic components cycling through the Solar System. He forced them to collide with and smash into Earth with enough force to upset equilibrium conditions. The violence of the impacts triggered great floods and fires that swept over the Earth's surface.

He stimulated some animal species to reproduce in enormous numbers and swarm over the Earth with yet other procedures. They consumed or destroyed Earthly resources that sustained other species. The swarms overwhelmed other species, indiscriminately causing their physical extinction.

He devised the testing procedures initially in the ideal Thought environment of his free, independent memory to achieve precise killing outcomes. Exercising his increasingly sophisticated free-Thought-processing capabilities, he sequenced Thought into structures that also described the real events and their real outcomes as the EOS reacted to the impact of his killing procedures or the release of his Atumic controls.

These Thought structures allowed him to identify those changes in ideal outcomes that the uncertainties in EOS utilities caused, and those that the Atumic actions of another god may cause. He compared the sets of outcomes. He focused only on anomalies. His Atumic caste automatically attributed any he was unable to match with those the EOS automatically produced to the actions of Thoth or his five siblings, using Atumic procedures to interfere with his absolute control of Earthly events.

Thoth was aware of Ra's intentions with the Earthly tests. His Thought processing was still much more sophisticated than Ra's, and produced Thought structures explaining the anomalies. The explanations focused on another abstraction the Atumic Ra had difficulty processing: that of the uncertainties inherent in the outcomes of COS and EOS utilities. These Thought-processing interactions increased Ra's unAtumic analytical skills.

The Atumic Ra continued to analyze Earthly events. He was unable to disassociate any outcome Thoth proposed from trickery. His analyses indicated that Earthly events were proceeding as he had designed. The real events he had triggered produced the mass killing outcomes on which he had focused. The autonomous EOS also reacted, as the Thought structures in his Atumic Thoughtbase confirmed, with automatic EOS procedures for establishing a new equilibrium in the altered state of the Earthly physical environment. His Atumic conditioning was able to determine that the new equilibrium was consistent with the outcomes of the Chaos-equilibrium recovery cycles inherent in the COS and EOS.

Ra returned his Atumic focus back to his responsibilities as supervisory god of the Solar System. To keep himself aware of events on Earth, he designed a security routine, a new utility that alerted him of any Atumic anomalies.

THOTH MENTORS HIS FIVE EARTHBOUND SIBLINGS AND BEGINS USING THOUGHT PROCESSING TO CONTROL EARTHLY EVENTS

The outcomes of Thoth's interactions with his elder brother Ra since the first traumatic events that followed his advent into the Solar System established his unAtumic Thought processing as his only godly survival tool. Thought processing in the privacy of his individual memory was his only absolute freedom and defence against his elder brother's corrupted combinations of Atumic/UnAtumic actions. Every action Ra initiated focused on ridding the Solar System of Thoth's presence—and now that of his five youngest siblings.

Thoth had made himself the ultimate Thought processor for his own survival. He focused on expanding his unAtumic free memory with Thought structures he retrieved from MEMORY and sorting them into immediately accessible Thoughtbases. He designed one Thoughtbase with Thought structures defining procedures he could transmit to Ra to defuse his constantly imminent irrational rage and divert him from actions whose outcomes could disintegrate the Solar System into Chaos.

Ra's increasing and inevitable sophistication in developing his unAtumic Thought processing promised many uncertainties and dangers for the Solar System. All were primarily the outcomes of his processing Thought structures simultaneously through his Atumic interface and in his free memory. The outcomes were always incompatible, triggering his confusion and incurring his wrath that he vented on his clones' progeny. He was forever to keep himself trapped in a network of Thought structures that always led to one outcome: the elimination of Thoth and their five siblings.

Thoth began mentoring his five siblings using his safest and most powerful survival tool, Thought processing. He had to direct them away from reacting as Atumic gods on finding themselves confined to the limitations of their Earthly environment. All five were legitimate cosmic gods with godly attributes and access to the Atumic Thoughtbase. He was uncertain how they would react to the physical restrictions that their HAP structures placed on their godly capabilities. They were all capable of retrieving Atumic Thought structures that specified the procedures to use as Atumic gods in reaction to events.

Thoth had to stop them from expressing Atumic frustrations with their limited physical capabilities. They were all capable of developing uncontrollable corrupted characteristics similar to those that Ra developed after Thoth's advent into the Solar System. In that event, Ra had only one cosmic god to contend with and now he had five more, all the direct result of Thoth's actions. One Atumic action from any one of the five would give Ra the Atumic right to appeal to Atum and insist on the banishment of Thoth and the five from the Solar System and into cosmic limbo.

Thoth's safest option was to co-opt their Thought processing as they entered the Earthly environment using the Thought communications capabilities he had developed while interacting with Ra. He transmitted two Thoughtbases to them. One contained Thought structures defining their capabilities as hominid animals and the dangers to their survival should they try to execute Atumic procedures to free themselves from their HAP restrictions. The other Thoughtbase contained Thought structures that he had retrieved from MEMORY to describe the EOS and the fixed set of utilities that established their physical environment. As each entered the Earthly environment, he stimulated their unAtumic communications channels into their free memory with Thought structures directing them to process his Thought structures and capture his messages before they accessed the Atumic Thoughtbase.

His Thought structures explained the conditions under which they could operate in the sanctuary of the Earthly environment. The IS components disseminated messages directing them to always comply with the survival procedures that Ra specified for the hominid animals. They were never to execute Atumic procedures. Ra had sequenced sets of Thought structures that specified directives for reacting to Earthly events. They had to obey the directives to avoid exposing themselves to Ra's wrath, but they were free to add sequences to Ra's procedures to improve their individual survival chances.

He described all the HAP capabilities and the advantages that hominid animals had over other animal species. He explained that hominid animals exhibited physical characteristics and capabilities that distinguished them from other Earthly species. They had opposable thumbs that gave them the capability to use their hands to execute physical survival procedures that few other animal species were capable of executing. The hominids were of a physical size and composition of cosmic and Earthly elements that allowed them to adapt quickly to changing environmental conditions. They could survive both on a smaller amount and greater variety of food for sustaining the efficiency of their HAPOS than other animal species needed to sustain the operation of their physical structures.

He explained one powerful attribute they had as the outcomes of godly sexual activities. They had access to MEMORY and the capability to retrieve free-Thought from MEMORY. They each had an open expandable individual memory in which to store the free-Thought and a Thought-processing facility functionally similar to those of their godly parents.

Hominids as with all animal species had a limited memory with a Thought-processing facility that allowed them to access only their closed physical survival Thoughtbases. The Atumic cloning procedures duplicated only the fixed Thought structures of the closed Thoughtbase that Atum designed to focus only on physical survival. Storytellers labelled the outcomes of this Thoughtbase as "instinct."

They had to react to these hominid instincts as did other hominids but their open individual memories gave them multidimensional freedoms to sequence Thought into structures with outcomes that only gods had the Atumic power to implement. The confines of their HAP structures and the physical bounds of the EOS limited them to the two- and three-dimensional freedoms that the EOS allowed. They could sequence Thought into structures to produce any outcomes in their individual memories, but the COS, the EOS, and their individual HAPOS limited them to implementing only a few.

He alerted them to the inevitability of the ultimate disintegration of their HAP structures from the EOS recycling and other utilities that controlled Earthly events. Most of the utilities maintained the equilibrium conditions in their operating environment and they had access to MEMORY for the Thought structures that specified utility routines and outcomes. Some utilities initiated events with potentially disastrous outcomes—massive Earthly upheavals that erupted from the Chaos at the Earth's core, overwhelming floods, and sweeping fires—that had killing outcomes.

He focused one set of Thought structures on describing the ultimate advantage their free-Thought-processing capability gave them over all other animals, including their hominid hosts. They were able to sequence Thought in the free space of their individual memories to describe the sequences in the Earthly physical events that the EOS utilities initiated. The capability allowed them to develop an independent survival Thoughtbase with Thought structures whose outcomes predicted the dangers that the utilities produced and their impacts on the operating conditions of their HAP structures. The pure hominids and other animal species simply reacted to each event, unable to connect the outcome of one event and the initiator of the next.

The new Thought structures specified procedures they could execute to avoid the dangers of Earthly events. The procedures allowed them to prepare for each stage in an event and avoid killing outcomes or damage to their HAP structures. As hominids they were able to demonstrate to the pure hominid members of their groups using primitive OralITIS and ImageITIS products to describe the new survival procedures and direct them to take shelter or move to safe locations before the events reached disastrous outcomes.

The outcome of these new procedures was an increase in the longevity of individual hominids that belonged to the groups to which the five Earthbound godly siblings attached themselves and an increase in their numbers. These outcomes prompted their hominid groups to treat the five Earthbound siblings Osiris, Isis, Seth, Nephthys, and Horus as their leaders and when their HAP structures finally disintegrated back to their cosmic and Earthly elements, as gods.

THE FIVE EARTHBOUND GODS ADAPT

The deadly Earthly events Ra had initiated triggered intense Thought processing in the five Earthbound godly siblings. The threats to the survival of their HAP structures as Earthbound godly entities were new ones for gods. They began to test the physical capabilities of the five hominid animal individuals in whose physical structures they chose to operate in the Earthly environment. They focused on sequencing Thought structures that described the limitations that the HAP structures and the EOS imposed on their godly capabilities.

The five Earthbound gods inherited both Atumic and unAtumic Thoughtbases, individual memories, and the free-Thought-processing capabilities of their parents. They had the freedom to access and process the Thought structures in their inherited Thoughtbases and to expand their individual unAtumic memories with a new Thoughtbase describing the threats to their HAP structures. They sequenced Thought structures with procedures that helped them limit damage to their HAP structures and avoid dangerous events.

They re-processed the rigid physical survival Thought structures of their hominid animal hosts. They identified the physical restrictions under which the hominid animals had to operate and that limited their implementation of the new survival procedures. They began to re-sequence the Thought structures to define procedures with which to adapt the physical capabilities of their animal structures and improve their chances of survival in the three-dimensional Earthly environment.

They accessed the Thoughtbase containing Thought structures defining the EOS procedures and outcomes. They processed the EOS in their free memories and defined the sequences that preceded and followed each stage of the events. They devised sequences that duplicated those of each stage and defined the outcomes that signalled the beginnings of each stage.

The outcomes of their analyses allowed them to define procedures that directed them to execute pre-emptive procedures protecting them from the final killing outcomes of fixed EOS utilities. The new procedures directed them to find caves in which the hominid group they joined protected themselves against the ravages of harsh weather conditions. They built safe havens to protect their groups against predator animal species. They adapted the EOS utilities for making fires: They reproduced the initiating conditions by striking two stones against each other to produce sparks to ignite dry leaves and other kindling in a process similar to that of Earthly lightning striking dry forest.

THE FIVE PRODUCE THE FIRST HOMO SAPS

The five Earthbound gods began developing a powerful feature of their free-Thought-processing capabilities. They began to identify outcomes they wanted to achieve. They sequenced Thought into structures specifying procedures that helped them achieve the outcomes within the restrictions of their fixed operating environment and with the Earthly resources available to them. Their analyses of the EOS indicated that many procedures allowed them some degree of freedom to duplicate the procedures and reach Earthly outcomes without disturbing the prevailing equilibrated environment excessively or provoking the EOS into executing procedures that produced violent reactionary outcomes.

Fully aware that they had to operate within the confining physical conditions of their HAP structures, they focused their Thought processing on describing all hominid animal characteristics. They separated those that could give hominids survival advantages over other animal species and those that allowed them to ameliorate the rigors of their physical environment.

Their physical structures, though puny and with physical strengths inferior to those of many other animal species, allowed the hominid animal group they joined to sustain their physical operating systems with smaller quantities and greater varieties of Earthly food resources than those other animal species required to sustain themselves. Most other animal species had to forage for and consume food resources constantly to sustain their individual operating systems.

The one set of physical activities the five Earthbound gods executed free of godly, clone, or EOS restrictions was the set defining the full range of unAtumic sexual characteristics. Ra's original clones and their sexual offspring stabilized with the imperative to execute only sexual reproduction procedures to sustain their numbers. They executed the sexual procedures only when their HAPOS conditioned the HAP structures of cloned females for reproduction.

The five Earthbound gods were free to execute their sexual procedures and express their sexual characteristics without restrictions. Their first implementations of their sexual procedures with hominid physical structures established a new Thoughtbase, with Thought structures describing the sexual outcomes.

Osiris and Isis joined as an exclusive sexual couple, as did Seth and Nephthys. Horus, without a complementary godly sexual partner, chose to execute his sexual procedures with female hominid animals who submitted to his sexual approaches.

They all produced sexual offspring, Horus more than the exclusive pairs of Osiris and Isis and Seth and Nephthys. The four elder siblings chose monogamous unions with their godly partners to initialize all their sexual characteristics.

Horus was free to execute his sexual procedures with many partners, but unlike his elder siblings, his sexual development was incomplete. The hominid females with whom he conjugated reacted with emotion only during their fertility cycles. In addition, the species survival imperative made sexual exclusivity for hominid animal females counterproductive. Life expectancy among hominid animals was uncertain and species survival forced the female hominid animals to make available their sexual features to any hominid male that was sexually ready to implement the physical act.

The offspring of the Earthbound gods stabilized physically as hominid animals. The sexual activities of the five had initially little impact on the hominid physical structures. Their godly-hominid offspring were physically similar to the offspring of Ra's clones, but they inherited some unAtumic characteristics from their godly parents. Most godly-hominid offspring inherited individual memories with capacities and processing facilities far greater than those of the pure hominid animals.

The characteristics of each offspring, godly-hominid, or pure hominid animal, were different enough and unique enough to identify each as a distinct individual. Each godly-hominid individually exhibited different capabilities for accessing, retrieving, and processing unAtumic Thought structures. As structural combinations of Ra's clone progeny and sexual gods, they responded to both Ra's controls and their unAtumic Thought processing.

The uncertainties of the EOS and godly sexual outcomes produced some godly-hominids whose clone characteristics dominated their reactions. Ra's Atumic conditioning inhibited them from accessing and processing Thought structures other than those in the fixed Atumic Thoughtbase defining their HAP survival procedures.

Other godly-hominid offspring stabilized with strong unAtumic characteristics. Their Atumic conditioning triggered only a cautionary emotion when they accessed and freely processed Thought in their free independent memories. Godly-hominid offspring exhibited wide variations of characteristics over many generations of reproduction cycles, all complex combinations of the characteristics of all their godly forebears.

Osiris and Isis passed on characteristics similar to those that dominated the activities of their godly parents, Geb and Nut. They focused their Earthly activities on completing the expression of all their sexual characteristics and initializing their pleasure emotion characteristics. They made them the characteristics whose expression they reproduced the most often. Their monogamous sexual relationship allowed them to fix many connections between sexual procedures and the pleasure emotions, extending the outcomes beyond the simple hominid animal reproduction demands of their HAP survival imperative.

Their frequent mutually-compatible sexual joining with intensifying pleasure outcomes established them as Earthly entities whose impacts on their hominid animal social groups were benign, nurturing, and calming. They transferred these dominant characteristics to their godly-hominid offspring, who in turn began to establish themselves as initiators of nurturing monogamous family units in the hominid animal collective.

Seth and Nephthys also freely executed their sexual procedures, but with less benign outcomes than those of Osiris and Isis. Seth had stabilized with cosmically-modified characteristics similar to those of his eldest brother Ra. In Seth, the uncertainties of cosmic and sexual outcomes had stabilized his unAtumic characteristics with weaker definitions than those of his godly brother Osiris. Atumic characteristics had a greater impact on Seth's ultimate character development.

Seth and Nephthys succeeded in initializing their unAtumic pleasure emotion characteristics, but in Seth the pleasure outcomes never matched the intensities that the unAtumic Thoughtbase in his individual memory described. This less-than-perfect godly outcome triggered his Atumic anger emotion and established a direct connection between his Atumic bias and his unAtumic sexual and pleasure characteristics. He attributed his Atumic anger to his godly confinement in the HAP structure, inhibiting him from executing his godly capabilities and achieving perfect godly outcomes. His reactions to events, Earthly or sexual, were inconsistent. Seth established himself as an Earthly entity whose activities produced disturbing and occasional destructive outcomes that made his hominid animal hosts fearful and wary of his presence among them.

Nephthys stabilized with most of the characteristics of her godly mother Nut, but with dominant ones establishing her as a defensive, sympathetic Earthly entity. As her godly mother Nut had executed procedures to ameliorate the outcome of her son Ra's Atumic anger, Nephthys executed procedures to ameliorate the impacts of Seth's Atumic anger. When Seth focused his Atumic anger on his peaceful godly brother Osiris, Nephthys helped her godly sister Isis hide. She protected Isis's offspring from Seth's unprovoked destructive actions.

Nephthys was unable to predict Seth's sudden angry outbursts to help him curb them. Their godly-hominid offspring inherited their characteristics; the sexual orientation of the offspring determined their dominant characteristics. The offspring with male orientations inherited more of Seth's characteristics, while those with female orientations inherited more of Nephthys'. Their male offspring tended to focus their anger on any impediment to their freedom to complete any Earthly activity they chose to initiate. They ignored any negative impacts of the outcomes on the survival or safety of other hominid individuals in their social groups.

Horus, the youngest of the five godly siblings, stabilized with a godly structure that exhibited a balance of parent characteristics. He inherited all the unAtumic characteristics of his godly parents, Geb and Nut. All the connections between the defining characteristic sequences fixed his godly structure in a uniform configuration, offering him uninhibited access to his unAtumic Thoughtbase.

He was free to express all his sexual characteristics and execute the full range of sexual procedures with female hominid animals. Without complementary sexual emotions to stimulate and guide his own, the female hominid animals allowed him to initialize only his unAtumic characteristic of contentment. They responded with enthusiasm only when their HAPOS conditioned their HAP structures to initialize the reproduction process for producing new individuals.

Both partners in a sexual coupling had to have access to the sequence specifying all pleasure emotion characteristics in their defining structures. The male and female characteristics complemented each other and each needed the other to trigger and complete the initialization. His female hominid animal partners were generally submissive and simply allowed him to use their complementary physical features to satisfy his sexual imperative.

His contentment, more lasting than the extreme sexual pleasure that kept his elder siblings absorbed in their mutual sexual activities, initialized a set of unAtumic characteristics whose expression focused on his relationship with the social group he joined. He initialized and reinforced his unAtumic emotion of pride with his successes in increasing the numbers of new individuals his frequent sexual couplings helped reproduce. His unAtumic pride in his godly-hominid offspring initialized a connection between the powerful Atumic survival imperative and the application of his godly capabilities to ensure the safety of offspring.

He began to focus his free-Thought-processing capabilities on sequencing Thought into structures whose procedures and outcomes gave extra protection from the killing outcomes of Earthly events to his godly-hominid offspring and their mothers. Most inherited his characteristics that focused on ensuring the survival of offspring and on the growth and survival of their groups.

They followed his lead in increasing the frequency of their sexual activities and directing their Thought-processing facility to focus on new and enhanced survival procedures. Over many generations his name carried and disseminated a powerful message.

Homo Saps who claimed descent from Horus named themselves Horus to disseminate the message that they, like their godly ancestor, focused on group survival. The Horus name identified them as leaders who devoted their Earthly activities and their free-Thought-processing capabilities to the production of outcomes that continually enhanced the survival and wellbeing of their social groups.

The godly-hominid offspring of all five Earthbound gods established the first Homo Saps, who exhibited a wide range of characteristics. The Homo Saps who owed their origin to the four elder siblings tended to classify themselves as more godly than hominid animal. They formed the first aristocrats. Future Homo Saps who claimed aristocratic status sequenced Thought structures proclaiming their advent as the result of a god conjugating directly with their Earthly mother or their Earthly father.

Horus' first offspring exhibited more hominid than Horus characteristics, but his characteristics were powerful. His prolific production of many generations of offspring out of many female hominids eventually produced many who had his characteristics dominating their Thought processing.

The Horus focus on Thought processing and individual memory development produced outcomes that demonstrated the power of Thought processing and memory as survival tools, tools that Thoth's communications with his five siblings and their offspring constantly promoted. Thought processing and memory development assumed the dominant role in directing the activities of Homo Saps. Their survival successes kept them expanding their Thought-processing facility and free memory capacities. Instead of binary Atumic killing procedures with their risks of physical damage to individual structures, they implemented and continued to develop applications that helped them achieve survival outcomes without killing or risk of damage.

The applications that had the most powerful impact on Homo Saps' development were those they made user-friendly to individuals. The user-friendliness allowed individuals to execute the applications independently to save themselves from the disastrous outcomes of the COS and EOS utilities. These successes initialized the Thought structure that disseminated the message that Homo Saps Thought processing was capable of overcoming or controlling COS and EOS events.

Some Homo Sap individuals and groups exhibited characteristics that limited their free-Thought processing or memory capacities. Some intentionally prohibited themselves from processing new Thought outside the set of fixed HAP survival Thoughtbase. Hominid animal species without Thought-processing capabilities succumbed quickly and easily to the killing outcomes of Earthly events. Storytellers labelled these Homo Saps as "fatalists" and the outcomes as "fate."

The HAP structures that the five Earthbound gods occupied were as vulnerable to physical disintegration from the cosmic and Earthly operating cycles as were all other hominid animals. They analyzed the impacts of these cycles on physical structures. The outcomes directed them to sequence procedures that allowed them to maintain and extend the operating viability of their HAP systems. These new procedures allowed them to make repairs to the damage that the cosmic and Earthly cycles of destruction and reconstruction automatically inflicted on their HAP structures. These repairs allowed them to operate for longer periods than pure hominid animals. Storytellers labelled the Homo Saps who designed Thoughtbases to contain these procedures and executed them as "healers."

The five godly siblings succeeded in extending the operating integrity of their hominid animal structures over many hundreds of generations, but only to delay the inevitable outcomes. The COS and EOS utilities disassembled critical sequences sustaining physical structures of all animals, causing them to collapse and disintegrate into their constituent cosmic and Earthly physical elements. To maintain Earthly equilibrium, the recycling EOS utilities re-sequenced the elements into new physical structures; some inanimate, others helping form and sustain the sexual offspring of animal species different from hominids.

The disintegration of HAP structures released the five Earthbound gods from their HAP confines. As independent godly entities with their unique characteristics, they entered a godly state that neither Thoth nor Nut predicted.

The Atumic prohibition on the execution of their godly capabilities and the birthing conditions that Thoth had set to satisfy Ra still confined them to the Earth environment. They operated as incorporeal entities in a Thought dimension possible only within the operating sphere of the EOS. Their only freedom was the freedom to transmit unAtumic Thought structures to their offspring on Earth. Storytellers labelled the five godly entities while they operated in this Thought dimension as "angels of the highest order."

Godly sexual activities with HAP structures that were themselves cosmically modified versions of Ra's clones, produced offspring with a wide range of godlike and clone-like characteristics. Some were more godlike than others and others were totally clone-like. These less-than-perfect godly entities entered the same Thought dimension as did Osiris, Isis, Seth, Nephthys, and Horus. Storytellers labelled these entities after their physical structures had disintegrated as "angels of lesser orders."

Some storytellers sequence Thought structures whose IS components disseminate messages that claim Homo Saps keep returning to the physical Earth dimension after their HAP structures have disintegrated. They label the repeating events "reincarnation." The storytellers claim that the cycles end when Homo Saps have achieved a Thought-processing capability that defines them as in a "state of grace," an ideal Thought dimension in which they overcome the terrors of their Earthly physical existence. Other storytellers sequence Thought structures disseminating the message that once the COS and EOS utilities disintegrate their HAP structures, they cease to exist in any form in any dimension.

THE NEW STATE OF THE SOLAR SYSTEM

The events that began with the cosmic god Nut's delivery of her first two sexual offspring established an Earthly hierarchy of godly entities. Except for Geb and Nut, all operated only in the Solar System. Nine gods now had at least some influence on events in the Solar System, primarily those in the Earthly environment that confined six of them: Ra, Thoth, Osiris, Isis, Seth, Nephthys, and Horus.

They exerted their influence mostly by using their unAtumic capabilities for transmitting Thought structures to each other and, eventually, to Homo Saps they spawned as outcomes of their godly sexual activities. Four gods: Geb, Nut, Ra, and Thoth, occupied the highest levels of the hierarchy and had the Atumic capabilities to execute procedures whose outcomes had the greatest influence on Solar System stability.

The five Earthbound godly siblings, Osiris, Isis, Seth, Nephthys, and Horus, occupied the next lower level. While they functioned in the Earthly environment in the HAP structures they had chosen, they occupied the lowest levels in the hierarchy. When the COS and EOS recycling utilities eventually disintegrated their HAP structures into the cosmic and Earthly elements, they functioned as incorporeal entities capable of operating only in the multidimensional extra-terrestrial environment. Storytellers labelled them "demi-gods" or "angels."

Other incorporeal entities—the spirits and avatars, all originally physical entities either as Ra's clones or as Homo Sap offspring to the five godly siblings—occupied the middle levels below the cosmic gods Ra and Thoth and the five godly siblings. All had some capabilities for sequencing and processing Thought structures, with procedures for influencing the stability of the Earthly environments. Ra had absolute control over all but two, Geb and Nut.

The corporeal Earthly entities, Homo Saps, and other animal species, occupied the lowest levels of the hierarchy. Some corporeal Earthly entities, primarily Homo Saps, had the Thought sequencing and processing capacity in individual memories to produce Thought structures specifying procedures that changed equilibrium conditions in the Earthly micro-environments in which the COS and EOS fixed and limited their operating freedoms to three-dimensional actions. Storytellers have persistently disseminated these messages.

Atum's threat extension effectively designated the sexual gods as a collective open to Atumic retribution for the aberrant actions of any one member of the group. The effect of this extension triggered Geb's awareness of his own vulnerability to the outcomes of events in the Solar System.

Geb—whose only function, so some storytellers have concluded, was to service Nut in her sexual needs—readied himself to intervene in countering the impacts of the interactions between Nut's seven godly offspring. Exhibiting a characteristic of male sexual gods, he was oblivious to the fact that he was as responsible for the production of the seven gods as was Nut. He was prepared to let Nut solve the problem as she had done before, but now his own survival was at risk. His readiness to act added more complications to the supervisory control of the Solar System. Two more gods, Geb and Nut, had now the Atumic potential to challenge or override Ra's actions.

Atum, whose simple Atumic analysis procedures were incapable of processing the complex connections between the outcomes of sexual gods' interactions, established the institutional practice of ignoring problems whose solutions were without clear binary outcomes and avoided analyzing outcomes of sexual events unless they reached the stage at which they caused more than minor anxieties in cosmic gods controlling adjacent cosmic systems. To avoid damage to their jurisdictions, the Atumic gods began to move their systems farther and farther away from the Solar System, constantly expanding the Cosmos.

Ra Chooses Fear as His Control Tool

Ra's interactions with his godly family had forced him to develop his unAtumic Thought processing to a level that was approaching that of his brother Thoth. He was able to impose some control on his irrational bouts of Atumic anger but his complex of interconnected Atumic and unAtumic characteristics still left him unpredictable. He had developed an unAtumic awareness that allowed him to monitor events on Earth automatically while focusing on his Atumic responsibilities.

The advent of his five godly siblings appeared to have made only one change: Their sexual activities, particularly those of his youngest brother, Horus, produced the new Homo Sap species of hominid animals. He was unable to distinguish them from his pure hominid clones. Their complex of characteristics—those they inherited from the sexual activities of his siblings with pure hominids mixed with those Atumic ones he embedded, unchangeable, as the core controls of the hominids—had little impact on their Earthly activities. All were functioning in accordance with his survival directives.

The one event he did note was the final disintegration of the HAP structures that his five siblings had originally invested. They had survived for many more hominid life cycles than his original hominids or their Homo Saps offspring had done. He had never determined how they had done so since they always followed his survival directives. They now functioned as incorporeal entities but still confined to the multidimensional Earth environment under his control.

An unusual event focused his awareness on Earth. He began to receive new Thought structures through his unAtumic communications channels. Only Thoth had transmitted Thought structures to him and his distrust immediately alerted him to the possibility of new Thoth tricks. His identification of the actual sources captured his Thought-processing focus: some Homo Saps exhibited strong, Ra-like characteristics.

He stored the new Thought structures in a new Thoughtbase in his free memory.

These new Homo Saps were transmitting with great intensity. The Thought structures contained sequences forming a mixture of messages that Ra sorted and began to analyze. One sequence described a fear of his godly wrath that matched his godly fear of Atum's retribution. Others described Ra as their father, their lord, and absolute controller of their Earthly survival.

The messages pleaded with him to use his godly power to tame the Earthly physical environment and establish physical conditions that allowed them to operate safely. They promised their absolute obedience to his directives and pleaded with him to end the disastrous killing outcomes of Earthly events. With every Earthly upheaval, flood, roar of thunder, and crack of lightning that the EOS utilities initiated, they intensified their pleas, attributing the continuing disasters and fearful events to expressions of his anger with their actions on Earth.

He reprocessed the Thought structures describing the reactions of the hominid animals. They offered him the same absolute obeisance as his Atumic conditioning forced him to give to Atum. These Homo Saps were acknowledging his godly supremacy.

He processed this Thought structure again and again. The outcome had initialized his unAtumic pleasure, a characteristic that he had yet to identify and express as part of his godly character definition. His analyses attributed pleasure as an outcome of expressing his sexual characteristics, but he had never been able to express them. He was without a compatible godly partner with whom to execute his sexual characteristics and his Atumic caste identified sexual activities as aberrant godly behavior. His attempts to use Atumic cloning as a substitute expression of at least his sexual reproductive characteristics had left him more frustrated than satisfied.

His Atumic caste compelled him to identify the source of the pleasure. Processing the pleasure Thought structures through his Atumic interface, he connected the pleasure outcome to the fear structure in his Atumic Thoughtbase. This Atumic conditioning allowed him to legitimize his expression and acceptance of unAtumic pleasure as Atumically compatible. Now all he had to do to stimulate his pleasure was to Atumically initiate an Earthly event whose outcome triggered the Homo Saps' fear of annihilation.

To test the reproducibility of his unAtumic pleasure, he devised new procedures to trigger minor Earth events. He brought the events to a quick end when he began to receive intensified fear messages from any group of Homo Saps. He began to sequence Thought structures specifying procedures Homo Saps should execute to earn his godly protection. With each event, he transmitted these Thought structures to the Homo Saps who were transmitting the Thought structures.

Most of the procedures required them to assume physical stances that exposed them to physical damage, as other animals did to declare their subservience to another animal. Storytellers labelled these procedures as "prayer rituals." When they did so, he immediately allowed the EOS to re-establish equilibrium or establish a new equilibrium for an Earthly environment whose physical conditions enhanced the survival chances and physical comfort of the Homo Saps who executed the prayer procedures.

He sequenced Thought structures describing every detail of their reactions and prayer activities. He sequenced Thought structures describing in equal detail the reactions of all other species his original clones had sexually reproduced. Only the Homo Saps species of hominids responded to him.

Most of the other animal species functioned as virtual automatons. Their activities focused on gathering and consuming Earthly resources to maintain the efficient operation of their individual animal OS and executing their sexual reproduction procedures. They simply succumbed to the damaging and killing outcomes of the EOS utilities without pleading with him to save them.

His pleasure came from prodding Homo Saps to express their fear for their HAP survival by triggering awesome displays of terrorizing Earthly events. Most Homo Sap groups prostrated themselves, sometimes mutilating their HAP structures, in elaborate rituals they designed to demonstrate their absolute submission to his control. A few expressed less than total submission, an anomaly that prodded his unAtumic curiosity.

The few groups who were less than totally submissive appeared to have been reproducing and surviving in numbers far greater than those of other groups. Their activities indicated that they were operating with some freedom from the godly fear and admonitions he had embedded in their core clone survival Thoughtbase. They executed the fixed set of animal survival routines, but they had added sequences to his HAP survival procedures to improve their outcomes. Their prayer rituals acknowledged his suzerainty as controlling god of the Solar System but they also directed some rituals to other gods. His analyses of their prayer rituals indicated that the other gods were his five Earthbound siblings, Osiris, Isis, Seth, Nephthys, and Horus, and Thoth, and his godly mother and father, Nut and Geb. These outcomes rekindled his distrust of Thoth's assurances that he, Ra, functioned as the only god in the Solar System.

He reprocessed his earlier analyses of Earthly events and began to sequence Thought into structures whose IS components speculated on how these Homo Saps had developed their advanced Thought-processing capabilities. This new development in his unAtumic Thought processing was a momentous one for Ra. His Atumic caste had always forced him to react instantly to outcomes only. He had never paused to speculate on the Thought processes that led to the outcomes or that the obvious outcomes could produce new, apparently unrelated outcomes such as those that followed Thoth's gaming outcomes. He had achieved a Thought-processing capability close to that of his brother Thoth.

The outcomes of his siblings' activities while they had operated in HAP structures, had indicated that they had complied with the prohibition against their using their Atumic capabilities. His analyses failed to identify the possible outcomes from their use of their unAtumic Thought-processing capabilities. With greater control of his anger, he used his own unAtumic Thought processing to determine how these Homo Sap offspring of his siblings achieved their Thought-processing capabilities in the confines of their HAP structures under his control.

The Thought processing of each Homo Sap generation expanded individual memories and processing capabilities. Each generation sequenced free-Thought into Thought structures with procedures that improved Ra's HAP survival procedures. Only under Thoth's tutelage could they have retrieved from MEMORY the EOS utility routines and reprocessed them in their individual memories to identify outcomes that modified their micro-environments to their survival advantage. Each outcome was within the stable range that the EOS permitted without disturbing the overall equilibrium.

Each generation inherited unAtumic memories from first, their godly, and then their Homo Sap progenitors. The uncertainties of the outcomes of sexual reproduction and EOS utilities conditioned their individual characteristics, giving some more complete memories and Thought-processing capabilities than others. They acknowledged their godly ancestors by soliciting from them Thought direction whenever their own Thought processing failed to produce successful survival outcomes.

Though the Thoughtbases they inherited identified them as facsimiles of their gods, most were extremely cautious in making their claims to godliness. Their inherited godly fear, both as Ra's physical clones fearful of physical destruction, and as gods fearful of godly banishment into a cosmic limbo that storytellers labelled as "Purgatory." Their fear kept them acknowledging Ra as their ultimate godly authority.

Their communications with him through their prayer rituals always led to agreements—storytellers labelled them "covenants"—in which Ra insisted on their acting as pure clones. He required them to accept and use only his survival procedures. He insisted that they confine their sexual procedures to those for reproduction only. As with all animal species, the Homo Sap females were fertile only for limited periods during their life cycles, the number of periods conforming only to the maintenance requirements of the hominid animal species in a sustainable Earth environment. They were producing offspring in numbers greater than that required for maintenance.

They were violating the terms of the covenant. Now they were praying for help and direction from other gods. His Atumic anger with them again began to grow in intensity, but he restrained himself to analyze the outcomes of his earlier expressions of his anger.

Before the advent of his five siblings, in the expression of his first anger reactions to events on Earth, Ra had acted as a pure Atumic god. The focus of his anger was his clone progeny. Their sexual reproduction activities, together with the impact of COS and EOS utilities, had produced entities that were extremely corrupted representations of his godly structure. Storytellers labelled them "abominations," brutes who functioned only to reproduce and consume Earthly resources.

His unAtumic analysis initialized a new emotion: disgust. Coupled with his Atumic anger mostly at himself for helping produce such ungodly caricatures—he had cloned them to be his images—his disgust had compelled him to eradicate them. His first weapons were small cosmic components orbiting his sun. He directed the cosmic components to collide with the much more massive Earth. The collisions caused cataclysmic physical upheavals on the Earth's surface. The outcomes changed survival conditions enough to cause massive annihilations of the abominations his cloning had reproduced.

He was unable to achieve complete success. Some of his clone progeny always survived. The uncertainties of sexual reproduction had produced species that thrived in the new physical environment and immediately began intense sexual reproduction activities, increasing their numbers in response to Earthly equilibrium requirements.

The only Atumic procedure that ensured complete success was the destruction of Earth. This outcome triggered his Atumic fear and he stopped himself from implementing the procedure. The destruction of Earth would destabilize the Solar System enough to cause severe disturbances in the surrounding stable Cosmos and invite Atum's retribution. Unable to let go of his Atumic anger, he directed himself to his unAtumic Thought-processing facility and began analyzing his options. His Atumic caste forced him to achieve a resolution to his anger.

That a few species of his clone progeny survived his destructive actions introduced more uncertainty in the absoluteness of his godly powers. His Atumic control had limits. He had to accept that the physical entities containing a facsimile of his Atumic structure were as much the outcomes of the EOS utilities impacting clone physical structures as they were of his Atumic cloning procedures. Their sexual orientations were also outcomes of COS and EOS activities, as were the sexual characteristics of his godly parents and grandparents. As Earthly physical entities, each species the clones formed operated independently, in accordance with the requirements of the micro-environments in which they operated.

The EOS made all his clone progeny, once they adapted to their Earthly limitations, necessary components in the maintenance of equilibrium conditions. His unAtumic Thought processing forced him to acknowledge his inability to eradicate them completely without also threatening his own survival in the outcome of the destruction of Earth.

His Atumic caste directing him again, he chose the one option his Thought processing offered. Instead of eradicating his clone progeny, he would control them with godly fear. His only safe control procedure, his individual safety his primary consideration, was to use his unAtumic Thought transmission procedures.

As he had done godly eons before, he accessed the procedures whose immediate outcomes eradicated masses of Earthly species. His ultimate intent was more subtle. He wanted to focus on Homo Saps. The annihilation of other species was a collateral outcome, an outcome that met only the single-valued procedural requirements of his Atumic caste. He added new sequences in this set of procedures. Their outcomes stamped him with new, twisted godly characteristics, neither Atumic nor unAtumic, but corruptions of both. The outcome demonstrated his characteristics for distorted but increasingly sophisticated Thought-processing capabilities that he kept private in his individual memory. Storytellers labelled the characteristics as "cruelty."

The messages he included in his transmitted Thought structures prodded Homo Saps' fear of his godly anger to intensities greater than their animal fear of annihilation. The messages threatened them with more gruesome outcomes both before and after he destroyed their HAP structures for ignoring his directives and paying homage to other gods. The messages contained instructions directing them to demonstrate, both in the Thought structures they transmitted and in their physical activities as animal species, their unequivocal acknowledgement of his absolute power over their survival and over other godly entities.

Some Homo Saps, more fearful than others, formed distinct social groupings. Each member of each group dedicated all their Thought processing exclusively to sequencing and transmitting Thought structures proclaiming their absolute obedience to his directives. To reward those who complied with his directives, he transmitted Thought structures directing them in procedures that helped them avoid the deadly outcomes of his cruel procedures. Homo Saps began to follow these groups as the Earthbound privileged executors of Ra's godly powers in ensuring their physical survival. Storytellers labelled them "priests."

Ra's initiation of new threatening and killing Earthly events served to demonstrate to Homo Saps his absolute control over their survival. They also demonstrated his suzerainty over Thoth and his five Earthbound siblings, whether they functioned as physical entities occupying HAP structures or as incorporeal entities unable to escape the extra-terrestrial, multi-dimensional Earth environment.

His control of Homo Saps was relatively easy. All he had to do was prod the powerful godly-fear emotion of the Atumic procedures embedded in their physical structures. The annihilation of large numbers of species and the messages he embedded in Thought structures he transmitted to them combined to make fear of him dominate their Thought processing. The messages forced them to blame their independent actions, taken without his sanction, as the causes of the Earthly catastrophes.

More and more Homo Saps responded to the calamitous events by falling to their knees to accept his retribution. They transmitted Thought structures with sequences acknowledging his supreme powers and with messages promising to abide by his godly directives.

Homo Saps' fear as animals overwhelmed the calming messages that the five Earthbound gods transmitted to them. His siblings could transmit procedures that ameliorated Earthly events, but the prohibition on their Atumic capabilities to change the course of events left them impotent.

Ra defined in greater and greater detail the limits to which he could manipulate the EOS without seriously impacting the Solar System or cosmic stability. One of his procedures focused on manipulating Earth's water component, as both a deadly weapon and a continuous survival threat.

Instead of bombarding Earth with cosmic projectiles, he changed the orbit of the moon—one of Thoth's sanctuaries—around Earth. He forced the moon into a degrading orbit, each revolution slowly bringing the massive Earth satellite closer and closer to Earth's surface. The change in the equilibrium keeping the moon and Earth subsystem in a stable configuration in turn initiated changes in survival conditions on Earth.

By focusing on the moon and Earth subsystem, Ra tested Thoth's Atumic neutrality and compliance with Atum's directives. He was challenging Thoth to counter his actions.

With each closing orbit of the moon, the Earth's bodies of water bodies—the oceans, seas, and large land-locked lakes—began to oscillate in higher and higher tidal cycles that engulfed coasts and shore lines with waves that reached deeper inland and devastated survival environments. The changing environmental conditions also produced downpours of rain, flooding the Earthly regions in which Homo Saps congregated and operated. The massive walls of water spread environmental havoc without discrimination. They engulfed and annihilated great masses of all Earthly species.

Ra could initiate procedures causing events that changed equilibrium conditions on Earth, but he had to leave the establishment of new equilibrium conditions to the EOS. The outcome that the EOS chose within the range possible was always uncertain. The cycles of actions and reactions could also cascade out of control, destabilizing the Solar System.

The EOS operated as a complex integrated set of utilities. Each produced different ranges of outcomes. Any one could arbitrarily trigger other utilities to produce sets of cascading events culminating in the destruction of the Earth-moon subsystem. He had to direct all his Atumic powers to preventing system disasters instead of monitoring the success of his strategies on the attitudes of Homo Saps toward him as their supreme god. Responding to his Atumic survival imperative, he stopped his actions and allowed the EOS to establish a new equilibrium in the Earth-Moon subsystem.

One outcome of his analysis helped defuse his Atumic frustrations with the limitations of his Atumic powers. Only *small* changes in the EOS could accomplish his objectives, so he began devising procedures that modified micro-environments on the Earth's surface. Homo Saps depended more on conditions in their immediate micro-environments than on the large environments remote from their safe havens and habitats.

Thoth's godly discipline in complying with the prohibitions on his Atumic actions also initialized a set of Thought structures with which he described his godly dilemma. His survival depended both on avoiding any interference with Ra's actions and on determining Ra's ultimate purpose. He identified possible outcomes, all devastating, all dependent on Ra's control of his complex of interacting Atumic and unAtumic characteristics.

Ra's initiation of the procedures for reducing the moon's orbit with Earth was for Thoth initially simply another indication of Ra's erratic Thought processing. He assumed that Ra had factored into the design of his destabilizing procedures all the uncertainties of the COS and EOS in reacting to the arbitrary, artificial instabilities his actions produced.

Ra's procedures forcing the moon into closer orbits around Earth increased exponentially the risks of the Solar System destabilizing. The minimum distance between cosmic components was critical to system stability. Thoth's dilemma was in his uncertainty about Ra having assessed the risks of his actions..

Thoth assessed his godly survival as much more precarious than that of the immediate aftermath of his advent in the Solar System. Then, Ra had reacted as a simple Atumic god; his declaration of his intent to eliminate Thoth was clear and open. Since then, his increasingly sophisticated but corrupted Thought-processing procedures that bypassed Atumic prohibitions allowed Ra to convince himself that the procedures were Atumically legitimate.

The EOS limited the physical reach of the massive tidal cycles and flooding. The tides and floods devastated only coastal regions and plains. Homo Saps moved to higher locations in mountainous regions of Earth or built floating sanctuaries in which they rode out the floods.

The high mountainous regions on Earth were hostile environments for most Homo Saps. Their Earthly physical structures operated with optimum efficiency in the temperate coastal and plains regions, where survival resources were abundant and most compatible with the maintenance requirements of their HAP operating systems.

HAP structures disintegrated rapidly in the new environments. The survivors modified survival procedures with sequences that allowed them to adjust to the hostile environments and neutralize the killing conditions. They built shelters that kept them safe from the sudden changes in Earthly environments. They brought with them other animal species that provided them with sustaining resources. Only those Homo Saps with the physical structures maintainable in hostile mountainous environments far from the floods survived the longest and reproduced offspring whose physical structures stabilized with the characteristics that enhanced their survival.

The changing environments had little impact on their sexual characteristics. The changing environments also had little impact on their Thought processing and memory facilities. Their Thought-processing imperative remained as powerful as their imperative for HAP survival. Each generation could access the Thoughtbases of their ancestors and reprocess the Thought structures for procedures adaptable to the changing environmental conditions. They began to build temples in the mountains high above the tidal cycles. They designed the temples as sanctuaries dedicated to praising Ra and acknowledging his suzerainty as the controlling god of the Solar System. They devoted themselves to executing the procedures of prayer rituals and transmitting Thought structures pleading for his mercy.

Ra returned the moon to a stable orbit and let the EOS utilities return the flooded Earth to a new equilibrium. Priestly Homo Saps in their temples high above the floods attributed the new equilibrium as his response to their pleas for his mercy.

They were wrong. Thoth had also been transmitting messages in Thought structures with IS components explaining the disastrous outcomes possible from Ra continuing to vent his Atumic anger both at Homo Saps for paying homage to their family of sexual gods, and at his siblings for producing Homo Saps. Thoth's messages forced

Ra to accept the fact that he was unable with all his godly powers to eliminate Homo Saps from the Earth without also endangering both their survivals as well as the survival of their godly family. But Ra's Thought processing had produced the same outcomes and Thoth's messages simply reinforced them.

Ra had achieved a Thought processing sophistication equal to that of Thoth's. His analyses of the outcomes of his actions that produced the Earth-wide flooding determined that Homo Saps' continued survival was primarily the outcome of Thought-processing capabilities that were close to godly. To ensure his continued suzerainty as the god of the Solar System, he had to control their Thought processing. He devised a new strategy, but one still based on their fear for their HAP survival.

His earlier analyses of the impact of his procedures for disrupting the Earthly environments in which Homo Saps operated indicated that the majority reacted with clone-like characteristics. Only a few controlled their fear and used their Thought-processing capabilities to produce Thought structures with procedures for ameliorating or avoiding the killing outcomes. He focused his Thought processing on the majority.

He reinforced his prohibition on their Thought processing outside the Thoughtbase he designed. He introduced a set of Thought structures with IS components disseminating new messages. The messages described those Homo Saps who independently produced new Thought structures with different survival procedures as the causes of deadly Earthly events. The messages directed the clone-like Homo Saps to forcibly subdue or kill any Homo Saps who exhibited free-Thought-processing capabilities that differed from those of the majority. In effect, he simply rekindled the original hominid animal antipathy toward the Thought processing of the first Homo Saps that his five siblings had produced. He chose as the Earthly disseminators of his new messages the priestly Homo Saps who had dedicated themselves to his service.

He supervised the first implementation of his new strategy. He reacted with fear prompting procedures to any Earthly event that produced an outcome different from those in his closed Thoughtbase. He triggered the EOS utilities to produce awesome displays of thunder, lightning, fires, Earth tremors, floods, and plagues in the micro-environment of the Homo Saps who initiated the events. He simultaneously transmitted to his Earthly representatives, the priestly Homo Saps, Thought structures whose IS components attributed the awesome events to his anger with Homo Saps for transgressing his rules for their Earthly operations.

His strategy was highly successful. Homo Saps complied with his rules for hundreds of generations, fearful of him and his priestly representatives. Claiming Ra's authority, they had formed themselves into a Homo Sap institution. Their godly authority—so they asserted— was for the benefit and safety of all Homo Saps and gave them the exclusive privilege of communicating directly with Ra for directives whose outcomes ensured their continued survival. All future Homo Sap institutions were to make the same assertion to justify their continued existence.

Ra directed his priestly representative to keep reinforcing the embedded fear in the hominid animal in all Homo Saps. He forced them to make the HAP survival imperative paramount, subduing any demands of their Homo Sap Thought-processing imperative for free-Thought-processing outside his closed Thoughtbase. He demonstrated fear as a powerful control tool, one that Homo Saps institutions were to use against any threat to their institutional survival.

Though Ra allowed the COS and EOS to return the Solar System and Earth environments to a new equilibrated state, Thoth still kept analyzing the possible actions his elder brother could take to rid the Solar System of him and his five siblings. He reprocessed his original analysis of Ra's character development, adding the outcomes of the great flood and other disasters Ra had initiated. The outcomes reinforced his original assessment that Ra could never entirely overcome his complex mix of Atumic and unAtumic characteristics, his uncontrolled expression of his Atumic anger and godly wrath always imminent.

A new outcome did indicate that Ra had fully developed his unAtumic Thought-processing capabilities and, more particularly, the feature that allowed him to sequence and process Thought structures in the privacy of his free memory without giving any overt indication that he was doing so. Ra had always made his emotional state obvious with some display of his Atumic capabilities for disturbing the stability of the Solar System. Those displays had given Thoth the advantage of determining his betting risk during their dice game. This new Ra appeared to be completely quiescent, a content and peaceful god. Thoth had to question this outcome.

Thoth monitored every Earthly physical event. His analysis indicated that the outcomes of the events were those that the EOS utilities produced in their continuous execution of the Earthly recycling procedures continuously destroying, reconstructing, and restabilizing the Earth environment.

He analyzed the Homo Saps' activities. They were exercising their animal sexual procedures without restraint but their numbers remained relatively stable. He analyzed this outcome. Their unrestricted sexual activities should have increased their numbers. His analysis identified one cause of this anomaly. As their numbers increased past a certain level, the EOS stabilizing utilities changed the Earthly micro-environment in which they operated. The changes reduced the availability of survival resources. Without adequate survival resources to maintain and sustain their HAPOS, the HAP structures of large numbers of individuals disintegrated back into the Earthly elements with which Ra had formed the clone ancestors. The outcome was a reduction in their numbers to levels that the available survival resources could sustain.

Thoth attributed another cause to the messages in the Thought structures that the members of the priestly institution were transmitting. The messages warned Homo Saps that Ra, the Sun god and their ultimate overlord, had forbidden all sexual activities except those that coincided with the single period in the Moon cycle during which the HAPOS conditioned female physical structures for the production and nurturing of new individuals. Yet, increasing their numbers was a survival imperative.

Thoth identified the only source of these anti-sexual Thought structure and messages: Ra, the only sexual god that insisted on imposing this restriction on Homo Saps sexual activities. His assessment of why Ra had assumed a quiescent Atumic character immediately changed. Ra had begun to use his Thought processing and unAtumic communications facilities to impose his Atumic will on Homo Saps. The outcomes were evident only in the Homo Saps' behavior. He was still primarily an Atumic god requiring absolute control over all events in the Solar System.

Homo Saps Reform Themselves

With the Earthly environment once again in an equilibrated state, the few surviving Homo Saps began to congregate around the priestly communities that had constructed their sanctuaries in the mountains above the flood levels. The priestly Homo Saps developed a rigid Thought-processing discipline with which they obeyed Ra's directives without deviation and processed only the Thought structures in his survival Thoughtbase. Ra punished Homo Saps for any deviation from his Thoughtbase with threatening Earthly events that the priests interpreted as evidence of his anger for their transgressions.

The priests developed the mountainous micro-environments to which Ra had directed them. They began to produce an abundance of survival resources. Homo Saps built their dwellings around the priestly sanctuaries. They made the temples the centers of their safe havens. They obeyed the directives the priests claimed were those that Ra had transmitted to them when they prostrated themselves in total subservience to him in the temples, acknowledging his absolute control over their survival and welfare. They focused their Thought processing on the Thought structures and messages the priests' constantly disseminated in OralITIS products containing their supplications for Ra's continued beneficence and protection from Earthly dangers.

Ra directed them to designate the temples as his private Earthly sacred houses and build them to function as ImageITIS products whose IS components constantly proclaimed Ra's eternal existence as their lord and master.

This use of an ITIS tool was one that Thoth had carefully avoided transmitting to Homo Saps when he began to help them develop their Thought-processing capabilities. Thoth had originally invented the ITIS tool of ImageITIS for Homo Saps to develop user-friendly applications for disseminating information about survival resources and conditions. Other Homo Saps captured the IS components of the ImageITIS products with the visual technologies embedded in their HAP structures. Thoth had also invented the ITIS tool of OralITIS for Homo Saps to develop survival applications for their embedded HAP aural technologies. That Ra had directed his priests to develop the ITIS tools as Thought-processing controls was, for Thoth, further evidence of his elder brother's complete but warped development of his Thought-processing capabilities. Ra's Atumic caste still forced him to reprocess unAtumic Thought structures through his Atumic interface to identify Atumically compatible outcomes.

The Thought-processing discipline and ITIS tool applications the priestly institutions implemented at Ra's direction helped them establish a stable environment for Homo Saps to develop after the flood and other disasters. This outcome presented Thoth with a problem. His analyses had indicated that Homo Saps had to develop a Thought-processing discipline if they were to develop beyond their HAP limitations. But they also had to have the freedom to sequence Thought into structures in the privacy of the individual memories outside the fixed boundaries of their embedded HAP survival Thoughtbase. The priestly institution had demonstrated the power of disciplined Thought processing in establishing stable social survival environments, but they were also inhibiting Homo Saps' free-Thought processing by stimulating their fear of Ra.

The solution to his problem was one that Homo Saps themselves developed as an outcome of the demands of the Thought-processing imperative. The outcome also demonstrated the limitations of the ITIS tools of OralITIS and ImageITIS.

Homo Saps grew their communities into large safe havens, centered around priestly sanctuaries and temples. As the communities grew, more and more individuals settled outside the perimeters of the safe havens far away from the priestly sanctuaries. They operated outside the range at which their HAP visual and aural technologies could capture the priestly ImageITIS and OralITIS Thought-processing controls. Without their constant processing of the inhibiting messages in the IS components of the priestly ITIS applications, their Thought-processing imperative compelled them to sequence and process new Thought structures in their individual memories.

Thoth left Homo Saps alone to develop themselves, but he kept monitoring their progress. Ra also left them alone because they still paid homage to him and were following his lead in establishing control institutions to help stabilize their social systems.

Many Homo Sap leaders, processing the Thought structures they inherited from their predecessors, used their ITIS tools for Thought-processing controls in the establishment of their Homo Sap civilizations. They determined that they needed to control at least four ITIS tools to stabilize their civilizations. Thoth helped them develop all four—

OralITIS, ImageITIS, CalendarITIS, and WritingITIS. Recognizing their power as control tools, they limited their use only to individuals and institutions they carefully selected.

Most Homo Saps used only the ITIS tool of OralITIS. They allowed their leaders and institutions to control the other ITIS tools as long as the outcomes succeeded in ensuring HAP survival. The leaders and their institutions began to lose control when Thoth invented the ITIS tool of WritingITIS and introduced individual Homo Saps to the specifications of WritingITIS products as stores of Thought structures emulating those of MEMORY.

Thoth left the development of the final ITIS tool—AlphabetITIS, that freed Homo Saps from godly and institutional Thought-processing controls—to the Homo Sap group storytellers, labelled as "Ancient Greek." Thoth was unable to influence the early civilizations in Ancient Egypt, Mesopotamia, India, and China; Ra had overwhelming control of their institutions, all of which operated in controlling Thoughtbases they designed to emulate Ra's.

5:

ITIS in Ancient Egypt

Dear Jim:

Those Homo Saps with classical training will probably question my crediting the Ancient Egyptians with major accomplishments in Homo Saps' development of ITIS tools. Storytellers who process Thought structures mainly from the Thoughtbase they label the "classical tradition" generally ignore the Ancient Egyptians and credit the Homo Saps who operated in Ancient Greece and Mesopotamia as Homo Saps' primary ITISanian ancestors.

But the Ancient Egyptians were the first to specify and develop versions of all five ITIS tools. Other Homo Sap groups developed versions of OralITIS, ImageITIS, WritingITIS, and AlphabetITIS, but failed to produce a consistent CalendarITIS.

The Ancient Egyptians identified Thoth as the godly inventor and disseminator of the ITIS tools. Even the Ancient Greek Homo Sap Socrates acknowledged that. Since Socrates was a primary sequencer and developer of classical-tradition Thought structures surely they must also. Yet the only message they capture from Socrates' Thought structures is an institutional one: the message that warns of the dangers to institutional controls of individual Homo Saps acquiring skills in the applications of WritingITIS. More importantly, the Ancient Egyptians demonstrated the good and bad impacts on individual Homo Saps when institutions maintain absolute control of the IT and IS components of the ITIS tools and their products.

Most storytellers have been fairly consistent in identifying the key Ancient Egyptian ITISans, but rarely as the original developers of ITIS tools. Most storytellers also operated in the classical-tradition Thoughtbase—their Thought structures were sequenced to support the presumptive classical-tradition outcomes identifying Ancient Greek or Ancient Roman Thought processing as the exclusive sources of Homo Saps' ITISanian developments. Their casual dismissal of the Ancient Egyptians—an ITISanian process they label variously "bias," "misinformation," or "disinformation"—prompted a few independent storytellers to begin processing Ancient Egyptian ITIS products and disseminating the messages, clearly identifying them as the initializers of the five ITIS tools.

They gave the name Narmer to the Homo Sap leader who developed ITIS tools and designed the operating utilities and applications with which he controlled and established the first Homo Sap civilization. They labelled as "Ancient Egypt" the Earthly region in which Narmer united all the Homo Sap groups which operated in the region and formed them into a civilization.

The IS components in the storytellers' Thought structures described Narmer as the developer of versions of the five ITIS tools and imposing institutional controls on their further development. Only then did he design and implement his operating system to establish the Ancient Egyptian civilization 40 Homo Sap generations (800 Earth years) before other Homo Sap leaders established different Homo Sap civilizations in other Earthly regions.

Unlike other Homo Sap leaders, Narmer gave himself the ITIS advantage before leading the military institution of his Homo Sap group in the hominid animal killing procedures whose outcomes helped him make his Homo Sap group the dominant group of all those he united. Narmer inherited from his predecessors the first versions of the three Ancient Egyptian ITIS tools of OralITIS, ImageITIS, and CalendarITIS. His predecessors had already specified and initialized the three sets of specifications.

His ITISanian processing of the messages in his predecessors' Thought structures described his job as the Homo Sap leader. His primary responsibility was to sequence Thought into structures specifying procedures whose outcomes ensured the HAP survival of his group—his tenure as Homo Sap leader depended on his successful fulfillment of that responsibility—but he was also to specify and develop the final two ITIS tools of WritingITIS and AlphabetITIS he needed to stabilize his civilization.

Further ITISanian Thought processing prompted him to sequence Thought structures specifying his need to integrate ITIS applications of all five ITIS tools as a suite defining the utilities of an operating system. The outcomes would allow him to transform the disparate Homo Sap groups he united into a stable civilization in which they operated as one group. His primary function as Homo Sap leader was to direct his Homo Saps in making the Homo Sap Thought-processing imperative dominant in directing their Earthly activities—at least as dominant as the HAP survival imperative had been in directing their Earthly survival activities in past generations.

The Homo Sap groups that Narmer united had already established their individual group operating environments at different locations adjacent to an Earthly feature that storytellers labelled the "Nile River." The Nile defined the Earthly region of Ancient Egypt and was the primary source of food survival resources that maintained the HAPOS that kept HAP structures in viable operating condition.

Before Narmer united them into a single group, they executed exclusively the directives of their HAP survival imperative. They competed with each other for control of the food survival resources, often executing hominid animal killing procedures whose outcomes annihilated whole groups of Homo Saps from the competition. The HAP survival imperative directed them to one set of outcomes: The control of enough survival resources to sustain their groups' HAP structures and the development of safe environments in which they were free of the immediate threat of physical damage from predator animal species.

The ITISanian Thoughtbase that Narmer stored and accessed in his individual memory was the cumulative outcome of thousands of generations of Homo Sap Thought processing—storytellers estimate 2,000 since Homo Saps established themselves as a distinct hominid species. Each generation transmitted the outcomes of their Thought processing to the next, primarily by developing new applications for their versions of OralITIS but also with ImageITIS products that illustrated survival messages and procedures.

Homo Saps focused their Earthly activities on the survival of their species, only occasionally activating the Thought processing and expandable memory features with which they distinguished themselves from other animal species.

As do all animal species, Homo Saps acquire the HAP survival Thoughtbase with procedures focused on activating their unique physical capabilities, each species with different survival mechanisms. They acquired the Thoughtbase as one of the outcomes of the animal sexual procedures that their parents executed in the brief periods when they freed themselves from immediate survival activities. All species reflexively access—storytellers label the action "instinctive"—their survival Thoughtbase for appropriate physical survival procedures.

Narmer initiated the first modifications to Homo Saps' exclusive dependency on the HAP survival imperative for direction of their Earthly activities. His skills with the five ITIS tools and the ITIS applications allowed him to develop safe and orderly environments in which Homo Saps could operate with a minimum of animal fear for their individual survival. He changed their absolute dependency on the HAP survival imperative to a co-dependency with the Homo Sap Thought-processing imperative.

The modifications he made forced his Homo Saps to rely only on him for the new Thought structures that would stimulate their individual Thought-processing imperative. The messages in the Thought structures directed them to begin focusing more on the stability and cohesion of his new civilization than on their individual survival needs.

Narmer acquired his original freedom to act as the principal Thought processor of his Homo Sap group directly from the members of his group. They chose him as the best individual to lead them. Homo Saps had implemented this leadership election process as a survival process. They directed the individuals they chose to freely sequence and process Thought structures in their individual memories, relieving the rest of the group to focus their Thought processing only on physical survival activities. In return, these individuals were to transmit their Thought structures to their groups during the dark periods when they rested. They made leaders of those individuals who developed successful procedures that enhanced group survival—and also the first storytellers. To fulfill his leadership function, Narmer's group freed him from most immediate survival activities and specified his primary group function as the sequencing of Thought into structures with procedures whose outcomes enhanced their physical survival.

Though Narmer had subdued the Homo Sap groups along the Nile River, some still resisted his authority. Accessing his ITISanian Thoughtbase for direction, he retrieved Thought structures that helped him develop a resolution to their resistance without executing hominid animal killing procedures. He declared himself a god-king, the Earthly representative of the Sun god Ra, and the first of the Ancient Egyptian civilization. He claimed the Sun god Ra directed him to make himself god-king.

Since his disparate groups operated in safe havens along the banks of the Nile and he was unable to personally deliver his directives in OralITIS products, he delegated the responsibility for imposing his Thought-processing controls to his priestly institution and the maintenance of physical order to his military institution. He claimed his establishment of the institutions as the outcomes of Thought structures and directives the Sun god Ra transmitted to him. Since his own sons would have godly connections through him, he appointed them as leaders of the two institutions.

Narmer's predecessors, all principal Thought processors of the ancestors of his Homo Sap group, had successfully specified, developed, and implemented the first three of the five primary ITIS tools, OralITIS, ImageITIS, and CalendarITIS. They had proven the value of their ITIS applications in enhancing the physical survival and growth of his Homo Sap group.

Narmer was unaware that both cosmic gods Ra and Thoth had sequenced and transmitted the Thought structures to him about the ITIS tool of AlphabetITIS. Thoth's Thought structures had explained the potential user-friendly features of the ITIS tool, features that would make WritingITIS products easy to produce and user-friendly to individual Homo Saps. This user-friendliness appealed to him since as the Homo Saps' leader, he was responsible for disseminating procedures that helped individual Homo Saps improve their Thought-processing capabilities.

Ra's Thought structures directed him to abort further development of AlphabetITIS even though he had already identified 24 symbols—ImageITIS products—that could function as the elements of AlphabetITIS. Ra's Atumic conditioning of his Thought processing had identified AlphabetITIS as a tool that undermined institutional controls of Thought processing in Homo Saps. AlphabetITIS would allow all Homo Saps to acquire individual skills in the applications of WritingITIS, skills that would allow them to develop their own individual Thoughtbases, bypassing Thought controls Ra and his god-king Earthbound surrogates broadcast to all Homo Saps. Narmer, his god-king requirements for control and stability dominant in the early development of his civilization, chose to follow Ra's directives.

Ra chose to broadcast his Thought structures aborting the implementation of AlphabetITIS. Thoth chose to remain quiescent, leaving Ra's Thought structures without opposition in Narmer's individual memory. As a cosmic god, Thoth had already determined the value of institutions in the development of Homo Sap civilizations. Institutions were necessary social components that would help Homo Saps initialize, develop, and stabilize their civilizations. He reserved sets of Thought structures he sequenced for transmission to a Homo Sap leader more advanced than Narmer. One set described the potential abuses of institutional controls on individual Thought-processing freedoms. Another described user-friendly AlphabetITIS as a powerful ITIS tool with which individual Homo Saps could develop and disseminate Thought structures challenging overbearing institutions.

He began developing a new Thoughtbase. He included all the Thought structures describing his predecessors' development of the ITIS tools of OralITIS, ImageITIS, and CalendarITIS. He added them to the Thought structures describing his development of the features and applications of the ITIS tool of WritingITIS. He included the set of Thought structures describing his development and aborting of the ITIS tool of AlphabetITIS. He integrated all ITISanian and HAP survival Thought structures and their outcomes in the new Thoughtbase. The outcome was a surprise. Applications and further development of the ITIS tools were more powerful than HAP killing procedures in their potential for maintaining the stability and longevity of his civilization. The one caveat was that he had to control their specifications and development and those who implemented them.

He began to analyze his control options. He had minimal control of both OralITIS and ImageITIS tools and products. Individual Homo Saps applied both ITIS tools primarily as HAP survival tools and had been doing so since their advent as a HAP species. They adapted the applications of the tools to the capabilities of the vocal, aural, and visual technologies embedded in their HAP structures. Individual Homo

Saps reflexively and instinctively activate crude applications of the tools whenever dangerous conditions threaten them.

His only control option was to force all his Homo Saps to adopt and condition their technologies to conform with the specifications of his versions of OralITIS and ImageITIS. He had, at least initially, full control of the specifications and applications of WritingITIS. He had complete control of his two CalendarITIS tools, both of which had helped his Homo Saps achieve the dominance they had.

ANCIENT EGYPTIAN CALENDARITIS

Narmer's predecessors had developed two versions of the CalendarITIS tool more than 50 Homo Sap generations before his. They developed both as information storage and retrieval systems. They controlled each with specifiable and consistent Thoughtbase management systems dedicated to retrieving procedures with outcomes they pre-defined. They designed the first to store and retrieve procedures for enhancing and optimizing their survival food resource production. They adapted the IT and IS components of the first to the outcomes they pre-defined for the second. They focused the ultimate outcomes of the second on controlling Thought processing in their Homo Sap groups with organization and regimentation features.

Storytellers labelled the first a "survival CalendarITIS." They labelled the second a "civic CalendarITIS."

His ITISanian analysis of the ITIS developments in the Homo Sap groups he had subjugated indicated that his CalendarITIS tools were unique in their specifications and consistency. His analysis also indicated that his CalendarITIS tools had to have some influence on his successes in making his group the dominant one among all of those that operated along the banks of Nile River.

Storytellers describe the Ancient Egyptian versions of CalendarITIS as singular Homo Sap constructs. The two tools were products of original Homo Sap Thought processing in individual Homo Sap memory.

Their development illustrated Homo Saps' progress in designing new applications for their Thought-processing capabilities. They had begun to specify outcomes and develop ITIS applications to achieve those outcomes. They had done so before, but the outcomes had focused on the simple binary, fight-or-flight outcomes of HAP survival. Those procedures anticipated and helped them avoid the killing or damaging outcomes of Earthly events that the COS and EOS utilities initiated. The CalendarITIS tools illustrated the beginning of Homo Saps' use of the cycles of the COS and EOS and repeating ImageITIS products to develop complex information storage and retrieval systems. The systems' designs helped them redirect their Thought processing to higher-level abstract outcomes than those of the simple HAP binary avoidance procedures.

Narmer's version of the civic CalendarITIS was one that the leaders of more than 50 Homo Sap generations of his Homo Sap group had processed and passed on to him in OralITIS products. The leader who initialized the first version specified a rigid design for the IT component. The Thought structures detailing the IT specifications also contained IS components disseminating unusual messages. The messages explained the need for the rigid IT components and directed future leaders to maintain them unchanged. The outcome of any change in the IT specifications would be the loss of the Thought-processing control feature of the civic CalendarITIS.

The messages explained that he designed the first implementation of the civic version to align with the phases of the survival version to ensure that his Homo Sap group accepted the civic version as they had the survival version. But, he warned, over the following generations the civic version would function out of phase—following a cycle that Storytellers label "Sothic"—with the survival version. The Homo Sap group's continued acceptance and execution of the routines of the civic version would indicate the success of the Thought-processing control feature.

The two CalendarITIS tools also established an abstract concept as a necessary feature for stabilizing and developing Homo Sap civilizations: Time. Systematized fixed cycles of Time ensured the orderly development of civilizations as long as institutions controlled the applications of Time. Individual Homo Saps instinctively organized their HAP survival activities around the abstraction of Time with their response to the control features of the cyclical light-and-dark periods of their operating environment. The CalendarITIS tools allowed their leaders to design artificial applications of systematized Time to control and coordinate both the HAP survival activities and Thought processing of large groups of Homo Saps.

THE SURVIVAL VERSION OF CALENDARITIS

Narmer's predecessors—their Homo Sap leadership responsibility to ensure adequate food survival resources foremost—had focused their Thought processing on the operation of the Nile River, their primary source of life-sustaining water. They sequenced Thought structures that focused only on the outcomes of the Nile events. They isolated in separate Thoughtbases Thought structures describing each of the Nile events. They arrived at a unique outcome after processing the Nile Thoughtbases: All Nile events repeated in cycles, the EOS controlling each to produce repeating and relatively consistent ranges of environmental conditions.

The EOS utilities controlled the volumes of water flowing in the Nile River channel, the speed of the flow, and the amount of Earthly material, silt, that the water carried in suspension. With each increase in water volume, the utilities increased the level of the water in the river channel and forced the silt-laden water to overflow the channel banks, flooding large flat areas of dry land adjacent to the river channel. On the flat land, the utilities reduced the speed of the flooding water to a level that lowered the capacity of the water to keep the silt particles in suspension and forced the silt to deposit from the water.

The deposit of the silt was the key outcome that ensured the food resource productivity of the land. Narmer's predecessors began to develop Thought structures that defined procedures for taking advantage of this outcome. Their analyses led them to the outcome that the effectiveness of the procedures depended on their success in predicting the Nile events. As they did with their development of procedures that anticipated and ameliorated future HAP survival dangers, they began to sequence Thought structures that identified and described the beginnings and endings of each phase of the Nile River operating cycle.

The outcome signalled a major advance in the leaders' Thought-processing capabilities. The outcome also initialized new Thought structures with outcomes leading to a similar major advance of individual Homo Saps' independent Thought processing.

As hominid animals, Homo Saps implemented their HAP survival activities to coincide with the COS's execution of utilities that controlled the light-and-dark periods of their Earthly environment. All Homo Saps, except their chosen leaders, limited their Thought processing to the procedures that ensured their survival needs in the Earthly conditions of each immediate light or dark period. Redirecting their processing on possible survival events of the next light or dark period distracted them from efficiently completing their immediate survival activities, the distraction often exposing them to physical dangers. But, for their leaders, predicting future Earthly events beyond those of immediate ones initiated a quantum expansion of their Thought-processing facilities and the storage capabilities of their individual Homo Sap memories.

Narmer initiated a critical memory expansion that led him to the final stage in the development of the survival CalendarITIS. He re-sequenced and interconnected Thought structures from those his predecessors transmitted to him in OralITIS and ImageITIS products, adding new sequences. The new sequences described the positioning and configurations of the cosmic components in cosmic space above his Earthly location. Storytellers labelled them "stars and planets."

His Thought processing directed him to define two outcomes. One was the periods of their movement across the sky or the periods between their appearance and disappearance. The other was to define their positions or configurations as ImageITIS products whose specifications coincided with the beginnings and endings of Nile River events.

He reprocessed the positioning Thought structures, reinforcing and confirming the stars and planets as components of a physical system with consistent periodic configurations that the unchanging COS controlled. He sequenced and initialized Thought structures that described the configurations as repeating in unique cycles: The COS utilities moved the cosmic components to return to the same relative positions in the Cosmos over definable numbers of light-and-dark periods.

The positioning cycle on which Narmer focused his Thought processing was that of the cosmic component that storytellers labelled "the star Sirius." The ImageITIS product he defined specified the relationship between the positioning cycle of the Sun and that of the star Sirius. After a set number of Sun cycles, the cosmic positioning utilities moved the star Sirius from a position below the eastern horizon of the Earthly location of the Nile River to a position above the eastern horizon, making Sirius the brightest star at that cosmic location. As an ImageITIS product, the star Sirius was user-friendly to the embedded visual technology of Homo Saps operating in their safe havens along the Nile River. They could identify the IT component easily and quickly. The cosmic utilities fixed the position of the star Sirius to coincide with the position of the Sun at the beginning of a light period when the Sun's light rays began to illuminate the Earthly region around the Nile River, prompting Homo Saps to begin their immediate survival activities for the light period.

He kept expanding the Thought structures to describe the cyclical positioning in Cosmos of other cosmic system components and groups of components that storytellers labelled "constellations." He developed a Thoughtbase of Thought structures describing the configurations, processing them only to meet his responsibilities as a Homo Sap leader. His Thought processing avoided any speculations on the functions of the constellations. His overriding survival function was to sequence Thought into structures defining procedures that enhanced the physical survival of his Homo Sap group.

He re-sequenced and processed Thought structures to identify relationships between the cycles of cosmic events and the cycles of Earthly events whose outcomes influenced survival environmental conditions in his location along the Nile River. In one Thought structure he identified the coincidence between two events. One was cosmic, the beginning of the Sirius star cycle. The other was Earthly, the beginning of the Nile River flooding cycle. Over several cycles of the two events, he confirmed the coincidence and added sequences to detail the specifications of both cycles and their beginning and ending markers.

The ImageITIS product he designed to specify the period of the critical cycle identified the beginning as the first appearance of the star Sirius above the eastern horizon of his Earthly location, just before the beginning of the light period of a minor Sun cycle. The star Sirius appeared at the same location on his eastern horizon at the beginning of each of the next 295 light periods. For the next 70 light periods the cosmic operating utilities moved the star below his eastern horizon, making the star Sirius invisible to his hominid animal visual technology.

Though he was unable to specify the star Sirius' exact position in the Cosmos during those 70 light periods when the star disappeared and normally he would assume the disappearance was evidence of the Cosmos destroying the star, he made another leap in Homo Sap Thought processing: He assumed it still existed and included the 70 light periods as part of the star's complete cycle. This event marked the initiation of a critical development in Homo Sap Thought processing. Their hominid animal Thought processing focused only on ImageITIS products that their HAP senses could capture and that had immediate impact only on their HAP survival. The new development initiated Thought processing applications that allowed Homo Saps to connect apparently disparate future events though they were unable to confirm the reality of the connections..

He designed the ImageITIS product describing the cyclical activity of the star Sirius to include a critical message focused on HAP survival. The message alerted him and his Homo Sap group to the fact that the reappearance of the star Sirius on the eastern horizon signalled the beginning of the light period during which the EOS forced the Nile River to begin the flooding cycle. This message also prompted him to begin organizing his Homo Saps for food resource production activities.

His successes in predicting the beginning of the Nile River flooding cycle prompted him to expand and fix the specifications of his survival CalendarITIS. He began to sequence Thought structures describing possible survival procedures associated with other markers of the Sirius star cycle. He began to expand the IT design of the ITIS tool. He anchored the design on the star Sirius' first appearance, then disappearance, from his eastern horizon. He established relationships between the IT features of the star Sirius cycle and the Nile River events.

He focused on the beginnings and endings of two primary events: the flooding cycle and the cycle of survival attributes with which the EOS conditioned the land after the water drained back into the river channel. He divided the primary IT of his survival CalendarITIS into 365 light periods. This IT specification emulated the periods of the complete star Sirius cycle and the primary Sun cycle..

He divided the 365 light periods of his survival CalendarITIS into three shorter periods: the first two of 129 light periods each and the third of 125 light periods. Storytellers labelled the three periods as "Akhet," the cycle of inundation; "Peret," the cycle of planting and growth; and "Shemu," the cycle of harvesting and storing of survival food resources. Each identified a macro period in which the EOS established different environmental conditions.

He sequenced sets of Thought structures defining procedures that took advantage of the macro conditions to enhance food resource production during each period. He stored the procedures in new Thoughtbases.

He captured and designed unique ImageITIS products that described micro-environmental conditions: The extent of the flooded areas, the quality of the top soil, the growth state and health of plants, and the readiness of the plants for harvesting. He designed the IS component as a Thought processing prompt. Each prompt directed him to retrieve from the new Thoughtbases sets of Thought structures, each set containing specific procedures that he designed to take advantage of the environmental conditions for optimum food resource production.

He designed an internet of Thoughtbases in his individual Homo Sap memory to identify the relationships between all IT and IS components of his survival CalendarITIS. He designed an operating system that controlled the implementation of any component of the ITIS tool. The operating system utilities automatically invoked other utilities or identified Thought structures directing him to new procedures in the food resource production applications. He designed the operating system to provide him with random access to any Thought structure in the internet of Thoughtbases.

The operating system responded to any Thought key associated with any feature of his survival CalendarITIS. To access the Thoughtbases, he had but to form in the free space of his individual Homo Sap memory any one of the unique ImageITIS products he associated with a set of environmental conditions. He designed a Thoughtbase management system that instantly connected his individual Thought-processing facility to the set of Thought structures defining the procedures he had to execute in any set of environmental conditions. He had but to sequence Thought structures describing the growth state of the food resource plants, the water level in the Nile River channel, the quality of the silt resource covering formerly flooded land, or other environmental conditions, and the OS instantly retrieved Thought structures offering him choices of activities that enhanced the production of food survival resources.

He designed the first application of his survival CalendarITIS tool to begin on the first light period marking the disappearance of the star Sirius from his eastern horizon. The first routine began the count in his individual Homo Sap memory of the light periods following the star Sirius' disappearance.

He designated a cadre of Homo Saps as exclusively responsible for the execution of the new survival procedures under his direction. He wanted to ensure that the only OralITIS products they processed were those he disseminated and authorized.

Most of the routines followed standard survival procedures that all Homo Saps, as hominid animals, automatically executed to meet their immediate survival requirements. Fulfilling his Homo Sap leadership responsibilities, he added new sequences with outcomes that improved their survival food resource productivity in each major and minor cycle of his survival CalendarITIS.

He embedded IS components in the new sequences with new messages. The messages explained the importance to their future survival of both the modifications to their standard procedures and the new procedures. He kept repeating the messages, exhorting his Homo Saps to embed them in their individual memories. Thought structures describing future survival were Thought structures that, in the past, Homo Saps deliberately avoided processing, the distractions diverting them from their need to constantly scan their environment for potential dangers. For hundreds of Homo Sap generations before Narmer's, all Homo Saps except those designated as central Thought processors confined their Thought processing rigidly to the HAPOS maintenance requirements, keeping their HAP structures in peak working condition during the specific light-and-dark periods in which they were operating.

They obeyed Narmer's directives instantly, implicitly trusting him as their chosen leader. All his past directives had always had outcomes that successfully enhanced their individual and group survival. They executed the new procedures, accepting the subtle but fundamental change in their Thought processing. They stored the Thought structures in an ancillary survival Thoughtbase since future survival procedures were incompatible with the immediate survival focus of their HAP survival Thoughtbase.

Narmer expanded the procedures with more specific directions as the first count of light period closed to the number of light periods between the disappearance and reappearance of the star Sirius, always inserting IS components with messages stressing the future survival advantages of their outcomes. The procedures directed his Homo Saps to begin preparing for the following cycles of his survival CalendarITIS. The messages always explained the connections between the activities that he directed them to complete and the impact of their outcomes on future survival food resource outcomes.

Their trust in his direction absolute, the cadres followed his instructions. They cleared the land of obstacles and debris that reduced planting and harvesting productivities. They levelled the land to slow the water movement and allow as much silt as possible to precipitate out of the floodwater before flowing back into the river channel at the end of the flooding cycle. He directed his group to vacate land and move their dwellings farther away from the river banks to increase the area of productive land.

He directed them to stay away from the flooded areas. He warned them of the dangers of drowning in the uncertain surges of water levels. He also warned them to beware of attacks from hippopotami and crocodiles that chose the Nile River as their operating environment and whose animal physical survival imperative identified other animal species they encountered in their watery domain as survival food resources.

He embedded sub-routines that emulated the cycles of Earthly events in the main application of his survival CalendarITIS. He designed each set of sub-routines to provide him with progressive counts of light periods from the light period he chose as the beginning of the three Earthly cycles of survival food production. He allowed the primary routine to simultaneously continue the controlling count of the light periods specifying the complete IT of the star Sirius cycle. He set the outcomes of the routines to highlight and prompt his Thought processing with ImageITIS products identifying the cosmic and Earthly physical markers that signalled the beginnings and endings of the repeating orders of cosmic and Earthly events.

From the outcomes of his first analysis of Thought structures describing cosmic and Earthly cycles, he had originally set the star Sirius cycle at 360 light and dark periods and the three Earthly cycles at 120 light periods each. With prompting from the cosmic god Thoth, who had added five days to the cycles to placate his godly brother Ra, further Thought processing directed him to refine these outcomes. He set the star Sirius cycle at a stable 365 light periods, but was unable to determine the precise periods of the three Earthly minor cycles. They oscillated around 120 light periods by a small varying number of light periods.

The outcomes of his subsequent analyses indicated that Earthly cycles contained many uncertainties. The Earthly operating utilities produced environmental conditions that differed from one cycle to the next but remained mostly within a range whose limits defined conditions beyond which HAP structures began to deteriorate. Though they were uncertain, he kept the Earthly cycles at 120 light periods to make the operation of his survival CalendarITIS more user-friendly to him and added an IS component identifying the uncertainties. At 120 light periods, the Earthly cycles melded seamlessly and uniformly into the controlling IT of the star Sirius cycle. He retained them as guideline IT markers with small variations to which he adapted his Thought processing as the light period counts approached 120.

INFORMATION STORAGE AND RETRIEVAL SYSTEM: INTELLIGENCE

He compiled all the Thoughtbases describing Nile events, macro- and micro-environmental conditions, and resource production procedures appropriate to each set of conditions in one internetwork of Thoughtbases. He designed a management system that formed the internetwork into a Thought storage and retrieval system. The management system allowed him to use Thought keys to access the information storage and retrieval system and retrieve the precise procedures appropriate to the food resource production requirements of each cycle or any stage in each cycle. Storytellers labelled the outcomes as "intelligence."

The intelligence outcomes prompted him to direct his designated cadre of Homo Saps in the orderly implementation of integrated suites of applications he designed to take advantage of specific environmental conditions that the COS and EOS utilities produced. These utilities executed automatically to produce a range of environmental conditions specifying the physical properties of the immediate environment. He used these properties as the operating parameters of the application with procedures that ensured optimum food resource productivity in the range of environmental conditions that allowed HAP structures to survive.

He sequenced the Thought structures he stored in his intelligence system to describe the details of the physical states—macro and micro—of the Earthly environment of the Nile River in every light period of the star Sirius cycle. As he processed them in his individual Homo Sap memory, the intelligence system's management routines compared the ImageITIS products that described micro conditions with those his HAP visual technology continuously captured. A match immediately directed him to the appropriate production procedures.

As the secondary count of 120 light periods after the flooding cycle closed, the intelligence system's management routines prompted him to retrieve Thought structures specifying the ImageITIS products describing the state of the environment most advantageous to the beginning of planting. The messages in the IS components prompted him to alert the designated cadre of Homo Saps to prepare for the planting procedures when the ImageITIS products he retrieved approached a match with those that his intelligence Thought structures described as optimum conditions for planting.

The intelligence system prompted him to follow similar procedures for the plant growth and harvesting cycles. The management routines kept him constantly forming ImageITIS products describing the healthy plant growth and the procedures for maintaining plant health for optimum productivity and readiness for harvesting.

The Earthly events and stability of the Nile River environment during his tenure as the Homo Saps' leader followed a regular pattern. The Sun's operating utilities produced consistent heating, lighting, and other environmental conditions without disastrous extremes that could impact HAP survival. The star Sirius consistently appeared on the eastern horizon on or close to the light period on which the Earthly operating utilities began the Nile River flooding cycle.

The Earthly cycles repeated without major changes in their periods or in the micro-environmental conditions. This steady state allowed his Homo Saps to produce a steady supply of food survival resources. Abundant food survival resources gave him more opportunities to free his Homo Sap Thought-processing imperative from the restrictions that his HAP survival imperative imposed on his Thought-processing freedom. He expanded still further his Thought-processing facility and the Thought storage capacity of his individual Homo Sap memory.

He redesigned his intelligence system as an open-ended one,

modifiable and expandable with further Thought processing in his individual Homo Sap memory. The uncertainties in the outcomes of the Earthly operating utilities that consistently set advantageous environmental conditions from one light period to another along the Nile River forced him to keep modifying the Thought structures or sequencing new ones, describing the different combinations of micro-conditions impacting the productivities of his Homo Sap group's food resource activities.

He modified the intelligence Thought structures with sequences detailing outcomes of micro-environmental changes, details describing the dryness of the soil, the intensity of the heat, the color of a leaf, and other ImageITIS products. He devised procedures whose outcomes ameliorated or enhanced the impact of any of the environmental conditions. His procedures emulated those of the Earthly operating utilities. They changed the utility outcomes, often accelerating or delaying their impact on the productivity of his survival food resource system.

His Homo Sap Thought-processing imperative—which some storytellers claim was the cosmic god Thoth—prompted him to reprocess these outcomes. He used them as starting nodes and began sequencing a unique set of Thought structures. Their IS components described his Thought-processing capability to change the uncertain, apparently arbitrary outcomes of Earthly operating utilities that he had accepted without question as immutable. Thought processing in the free space of his individual memory gave him the freedom to overcome the limitations of utility outcomes and modify them to improve still further food resource productivity. He could overcome the shortage of water by artificially pulling water from the Nile River and irrigating the land. He could overcome the problems of seeding hard-pan surfaces by ploughing and loosening the soil.

Processing this unique set of Thought structures he identified a unique message: His Thought processing gave him—a Homo Sap, a Thought-processing animal—the godlike capabilities to change his Earthly environment. He had the freedom to intervene in the operation of the EOS and manipulate the Earthly environment to suit his purposes. He was to make himself god-king on the assumption that he was the only individual with these godly Thought-processing capabilities; an assumption that later Homo Saps were to refute.

UNEXPECTED OUTCOMES

The success of Narmer's survival CalendarITIS and the resulting abundance and free distribution of survival food resources initiated events he failed to anticipate. Their outcomes left him with less than absolute control over the development of his Ancient Egyptian civilization. His analysis of the events—specifically the increasing numbers of his Homo Saps and their operations in safe havens outside his immediate control—directed him to begin developing the specifications of a supervisory control system that allowed him to re-establish his absolute control over the activities and Thought processing of Homo Saps who operated in the safe havens.

The direct supervision of the applications of his survival CalendarITIS left him with few opportunities to attend to his other Homo Saps leadership responsibilities. He stored all the production system specifications in his individual Homo Sap memory. He had to transfer the specifications to other members of the cadre responsible for executing the survival food resource production procedures in order to free himself to attend to his other responsibilities.

The only universal user-friendly ITIS tools the members of his cadre were capable of using to capture new Thought structures were the Ancient Egyptian versions of OralITIS and ImageITIS. They had already acquired skills in the use of these two ITIS tools in capturing HAP survival applications. They automatically stored any procedure that enhanced their HAP survival in their individual Homo Sap memories for instant retrieval.

He began to develop and disseminate OralITIS and ImageITIS products describing all the features of his survival CalendarITIS to his food production cadre. Though his analysis of possible outcomes indicated that one was the loss of his absolute control over the development of the ITIS tool, he was careful to include IS components with messages declaring the specifications as those he with his godlike capabilities established and only he could change them.

The messages had their intended effect and most of the members of the cadre followed the specifications and procedures without deviation. A few individuals processed and reprocessed the new Thought structures. Their Thought-processing imperative directing them, they began to analyze and emulate the Thought-processing procedures Narmer had developed to arrive at the production system. His messages powerful admonitions in their individual memories, they focused their Thought processing to operate within the bounds of his specifications. They focused on reprocessing and expanding the Thought structures defining the sets of procedures that they were individually responsible for implementing. They made the procedures more user-friendly for them to execute, adding sequences that enhanced their outcomes and improved still further the productivity of the food resource production system.

Narmer was unable to admonish the few for their independent Thought processing in individual Homo Sap memories. They had operated within the specifications of his survival CalendarITIS and increased the output of food resources. Their independent Thought processing had, in effect, enhanced his reputation as Homo Saps leader. But in failing to admonish them for independent Thought processing and modifying his initial production procedures, he tacitly gave permission to other individuals to free their Thought-processing imperative from his restrictions.

This event was critical. Successes in improving the productivity of individual activities prompted some individuals to begin describing their individual Thought processing as godlike.

Narmer's assignment of survival food production responsibilities to a small cadre who acquired specialist skills left many Homo Saps operating in his safe havens without the need to constantly forage for survival food resources. The safe havens also freed them from the need to ensure their HAP safety from predators, the responsibilities of another specialist group, members of his military institution. Most continued to confine their Thought processing to their HAP survival Thoughtbase, their primary responsibility. Processing Thought outside this Thoughtbase was Narmer's responsibility. They retrieved Thought structures defining survival procedures less immediate than food gathering and defence.

Their physical safety in the safe havens allowed them to execute one set of survival procedures with complete freedom. They began to execute their sexual procedures. Unlike most animal species, Homo Saps executed their sexual procedures at every opportunity their freedoms from other physical activities offered them.

The sexual driver in the OS of most animal species was Ra's core spiral structure that he duplicated using Atumic cloning procedures to form the progenitors of each Earthly species. The limitations of the Atumic cloning procedures and the impact of COS and EOS utilities modifying progenitor structures to make them compatible with Earthly operating environments in which Ra first formed them restricted most to execution of reproduction procedures only. The EOS modifications restricted the animal OS's activation of the sexual drivers to periodic events. The Earthly environmental factors determining the frequency of the events were the EOS's sustainability requirements—the need to maintain the balance between Earthly resources and consumers in constantly changing and uncertain Earthly operating environments

The EOS modifications set the beginning signals of each reproductive event in both female and male HAPOS. The EOS utilities activated female HAPOS utilities to initiate the production of visual and olfactory ITIS products to broadcast the beginning of an event. The male HAPOS captured the ITIS products through the males' embedded visual and olfactory technologies, alerting them to the productive condition of the females.

The IS components contained a message for the males: permission to approach the female to execute and complete the sexual procedure. Processing the message also automatically activates driving utilities that prepare male sexual technologies to facilitate the execution of the sequence of activities whose ultimate outcome is the production of new individuals for their species.

The EOS establishes different frequencies for different species to independently control the sustainable numbers of each species. Any changes in these numbers above sustainable levels threaten the species with extinction from a shortage of survival resources capable of sustaining the numbers. The lack of survival resources maintaining the efficient operation of the male and female reproductive systems might also force the animal operating systems to inhibit the activation of sexual technologies. The outcomes automatically reduce the numbers in the species to sustainable levels and allow the EOS to restore Earthly equilibrium conditions.

The HAP substrates on which Homo Sap Thought processing facilities operate also respond to the EOS equilibrating requirements. Homo Saps automatically function as animals, without Thought processing. Homo Sap females periodically produce ITIS products similar to those of other animals to signal the readiness of their HAP structures for the production of new hominid individual, but the ITIS products are less obvious and definitive than those of other animal species. Homo Saps' physical senses operate less efficiently than those of other animals. When they function as pure animals—technical components of the Earthly environment, responsive only to EOS equilibrium requirements—Homo Saps reproduce only enough to sustain themselves as a species. But, they are one of the most prolific of animal species.

The sources of their prolific reproductive capabilities are godly, one inhibiting and the other without restrictions. Animals respond only to the single sexual procedure that Ra, their primary godly progenitor and director of their Earthly activities, restricts them to. Homo Saps, when functioning as pure hominid animals, attempt to execute only this single sexual procedure, responding to Ra's directive. Their uncertainty lies in determining exactly when the female HAPOS conditions their structures for reproduction of new individuals.

As Thought processors they have access to the complex set of sexual procedures they inherited from their secondary godly progenitor, the Earthbound god Horus. Horus made his advent in the Earthly environment with restrictions only on his execution of Atumic procedures. He was free to execute the complete set of sexual procedures with hominid females as often as he chose. His offspring—the first Homo Saps—inherited this freedom and the complete set of unAtumic unrestricted sexual procedures.

The outcome of Homo Saps' freed sexual imperative was a quick increase in the numbers of Homo Saps who operated in the safe havens and in the expansion of the safe havens to accommodate the increased numbers. This outcome also presented Narmer with supervisory control problems.

Homo Saps' safety in the safe havens also allowed the HAP survival imperative to execute survival procedures that increased still further the surviving numbers of individuals. The ImageITIS and OralITIS products that Homo Saps captured while they operated in the safe havens described immediate physical environments free of threats from predator animal species. Without immediate threats, the HAP survival imperative automatically processed specific sets of survival Thought structures from their HAP survival Thoughtbase. The Thought structures specified procedures directing HAPOS to place animal physical structures into physically quiescent states, an operating mode that signals the HAPOS to activate specific system drivers.

The drivers call HAPOS subsystems whose procedures focus on repairing damaged elements of HAP structures and re-establish optimum HAPOS conditions. The subsystems operate automatically, as long as the HAP structures remain in the quiescent maintenance state. Without the need to maintain a constant alert, the HAP survival imperative releases Homo Saps from the restrictions on processing other than immediate survival Thought structures.

Many Homo Saps relieved from immediate survival activities, react as do other animal species and simply keep their HAP structures inactive. They rest or sleep. They ignore the demands of their Homo Sap Thought-processing imperative, and process only those survival Thought structures that direct them to rest or sleep, and wait for their leaders to provide them with Thought structures directing them to undertake other activities. Their resting and sleeping allow the HAPOS maintenance utilities to complete repairs to HAP structures, returning them to optimum operating conditions. The outcomes increase again the number of viable individuals in the safe heavens.

INDEPENDENT THOUGHT PROCESSING

Some individuals, who operated outside the aural or visual range of their leaders' OralITIS and ImageITIS products that directed Homo Sap activities, responded to the demands of their Homo Sap Thought-processing imperative. They began exercising their Thought-processing facility in the free space and privacy of the individual Homo Sap memories. The outcomes they produced differed from those of their Homo Sap leaders. Without their leaders' ITIS products constantly prompting them, they began to re-sequence and reprocess their leaders' Thought structures. They began to modify and execute procedures whose outcomes often disrupted the social order and opposed the outcomes that their Homo Sap leaders imposed on other members of their groups.

NARMER BEGINS TO ESTABLISH ITIS CONTROLS ON THOUGHT PROCESSING

Narmer began processing Thought structures identifying the problems he faced in controlling the coherence and stability of his Ancient Egyptian Homo Sap civilization along the length of the Nile River. He produced two outcomes. One described his primary control tool as his military institution. He could locate contingents of his military institution in all his safe havens and use them to threaten all Homo Saps who opposed his directives or disrupted the social order with the destruction of their HAP structures. He had used the hominid animal killing procedures to subdue all the Homo Sap groups along the Nile but some groups still opposed his authority. They submitted to his authority only from their hominid animal fear of his killing procedures.

His analysis of his original successes identified the source of those successes. The success of his food production system was an outcome of the coordinated activities of each member of his food production cadre. They all processed the same Thought structures, all of which he had sequenced and transmitted to them in OralITIS or ImageITIS products. Their continued obedience to his directives rested on his continued control of their Thought processing.

The increased opportunities for free, unrestricted Thought processing in the privacy of individual Homo Sap memories exposed the technical inefficiencies of his ITIS tools. He initially designed his ITIS products to be most effective in directing and controlling the activities of the smaller group over which he began his leadership tenure. His first OralITIS and ImageITIS products allowed him to deliver the Thought structures containing his directives directly to each individual Homo Sap as the group gathered around him. Each simultaneously stored exactly the same OralITIS products in individual Homo Sap memories. When they executed the procedures or directives in the ITIS products, they processed the ImageITIS products describing Narmer as he was delivering the Thought structures. They undertook the activities with his voice in their individual memories repeating the instructions as they began and never deviated. They all acted in unison.

To accommodate the larger group he expanded his safe havens and allowed more and more individuals to function outside his direct supervisory control. They operated at distances farther than the receiving range of hominid animal aural and visual technologies for their effective simultaneous capture and storage of his OralITIS and ImageITIS products. His survival CalendarITIS product focused only on the efficient organization of the cadre of Homo Saps he directed to dedicate their primary group survival activities to the efficient operation of his survival food resource system.

He reprocessed the Thought structures describing the features and applications of the three ITIS tools, OralITIS, ImageITIS, and CalendarITIS. He focused on their adaptability to his leadership requirements for disseminating control Thought structures to larger groups of Homo Saps, many of whom operated outside his direct supervisory control space. The IT of OralITIS and ImageITIS products allowed him some design flexibility, but the physical limits of the operating ranges of hominid animal embedded oral, aural, and visual technologies, limited the effectiveness of their applications for disseminating Thought structures to large Homo Sap groups.

Homo Sap groups Narmer made part of his Ancient Egyptian civilization processed Thought structures from Thoughtbases with outcomes different from those he had developed to direct his Homo Sap group. The ITs they specified for their OralITIS and ImageITIS products were often incompatible with that of Narmer's group. They had adapted their embedded aural and visual technologies to capture, process and react to different communications protocols. They sequenced their Thought structures and outcomes to fit their IT specifications, restricting their ability to easily capture the messages in the IS components of Narmer's ITIS products. The outcomes produced confusion and often violent reactions between the groups.

His ITISanian analysis indicated that his only option was to impose his versions of OralITIS and ImageITIS on all the groups. He implemented this option, but his ITISanian analysis also indicated that this option was only the first stage in consolidating his civilization. The new groups could simply acquire skills in the applications of his ITIS tools for their own survival. They would still operate independently as groups and continue to use their own Thoughtbases and versions of ITIS tool within their groups. He needed to control Thought processing in each individual Homo Sap memory to complete the integration of all groups into one group of Ancient Egyptian Homo Saps.

Unlike the ITIS tools of OralITIS and ImageITIS, Narmer had

devised, initialized, and fixed independently the design of his survival CalendarITIS. The outcome was a unique construct of his individual Thought processing in the free space of his individual Homo Sap memory. Together with the food production Thoughtbases he stored in his individual Homo Sap memory, his survival CalendarITIS was user-friendly and a true information storage and retrieval system. His food production cadre quickly captured the controlling Thoughtbases he transmitted to them in OralITIS and ImageITIS products as he demonstrated the procedures. They had made themselves experts in the execution of the food production procedures, some actually improving the procedures and expanding their individual Thought-processing facility.

He had operated in his Homo Sap leadership Thoughtbase to develop his survival CalendarITIS. He focused only on leadership outcomes that improved the safety, survival resources, and operations of his Homo Saps.

Still operating in his leadership Thoughtbase, his analysis of the new version of CalendarITIS he had begun outlining in his individual Homo Sap memory identified features whose outcomes conflicted with those of Homo Sap leadership responsibilities. In order to stabilize and develop a coherent Homo Sap civilization from disparate Homo Sap groups, he had to ensure that all, wherever they operated along the Nile River, processed Thought structures with the same messages and outcomes and automatically acted in concert. To achieve that outcome he had to function as a god-king, operating exclusively in his god-king Thoughtbase, focused on absolute control of both physical and Thought processing activities in all his Homo Sap groups and safe havens. The outcome would inhibit free individual Thought processing, a factor that allowed his food production cadre to make the survival CalendarITIS successful.

To help him reach a resolution of the conflict, he implemented an ITISanian procedure that had allowed him to resolve other conflicts. He exited all Thoughtbases, expanded his Homo Sap memory capacity, and began to operate in the free space, free of the directed outcomes of the Thoughtbases he had designed. The freedom allowed the ITISanian procedure to access MEMORY and sequence Thought into structures whose IS components and outcomes offered alternative resolutions to the conflict.

One set of Thought structures he kept returning to, and reprocessing contained IS components and outcomes indicating that imposing Thought-processing controls as god-king would ensure greater stability in his civilization and result in greater safety for his Homo Saps. The greater stability would eventually allow individuals with more opportunities for them to process free-Thought in the privacy of their individual Homo Sap memories. The IS components disseminated the message that individual Homo Saps would free themselves from Thought-processing restrictions despite their fear of his god-king retribution. This was an ominous message for his god-king status, one he included in his god-king Thoughtbase as an alert to his successors to whom he would pass his Thoughtbases.

Another outcome indicated that the success of this resolution depended on his successors' voluntarily lifting the god-king restrictions on Thought processing as soon as they had established a coherent and stable state in his new civilization. This was only one aspect of the conflict. He was unable to sequence a Thought structure that defined the state in which to lift the restrictions. He had to rely on the Thought-processing integrity of his successors to keep assessing the stability in each generation. But they were Homo Saps and forever open to the Thought processing direction of the cosmic gods Ra and Thoth, who could override any warning or alert he passed on to his successors..

His ITISanian analysis of all the possible outcomes indicated that the final outcome depended on the resolution of the conflict between Ra and Thoth. The resolution of that godly conflict exceeded his powers as their Earthly representative. He chose to establish stability as a first priority and leave the ultimate outcomes to Homo Saps' Thought processing.

THE INITIALIZATION OF THE CIVIC VERSION OF THE ANCIENT EGYPTIAN CalendarITIS

Narmer began the initialization of the new CalendarITIS with the outcomes of his ITISanian analysis of the successful features of his survival CalendarITIS. He reprocessed the Thought structures describing the design and attributes of his survival CalendarITIS. He disassembled the Thought structures, separating the sequences describing the IT of his survival CalendarITIS from the IS components. He stored the sequences describing the IT in a separate Thoughtbase.

He reprocessed the IS components and deleted all messages connecting the ITIS tool with his food survival resource production system.

He analyzed the new Thoughtbase. One set of Thought structures described the IT components of a template for the ITIS tool of CalendarITIS. He could adapt the template to any version of CalendarITIS he could devise. The IS components described two features on which he could develop applications for organizing and regimenting large groups of Homo Saps to execute procedures and achieve outcomes that he disseminated to them.

He identified the key control attributes of the template design as the predictability and regularity of the IT defining Earthly and cosmic cycles. The EOS and COS produced immutable ImageITIS products that signalled the beginnings, endings, and different phases of consistent repeating events. Those attributes and features allowed him to design, embed, or remove IS components from the Thought structures specifying a version of the ITIS tool without corrupting the tool's organizational and regimentation controls. Storytellers labelled the new version as the "civic version" of the Ancient Egyptian CalendarITIS tool.

He designed the IT of his civic version of CalendarITIS to emulate the IT of cosmic and Earthly cycles as he had done for his survival CalendarITIS. He chose ImageITIS products user-friendly to Homo Saps.

From their advent as a hominid animal species, Homo Saps had adopted the ImageITIS markers of the cosmic and Earthly operating cycles as user-friendly ITIS products for regulating their Earthly activities. He chose as the controlling IT of his civic CalendarITIS the three inter-related cosmic cycles through which the COS moved three cosmic components, the Sun, the moon, and the Earth. The cosmic operating utilities repeatedly fixed their positions relative to each other in the same locations in Cosmos over definable repeating periods.

The IT of these cycles was the most user-friendly for Homo Saps. All Homo Saps captured and processed ImageITIS products of the Sun, Moon, and Earth as part of their HAP survival procedures. They adapted the ImageITIS products as prompts that activated the drivers of HAPOS utilities, maintaining individual HAP structures in optimum operating condition.

They used the ImageITIS of the Sun at the beginning of each light period to begin their food gathering and other survival activities. They used the ImageITIS of the Sun disappearing behind their western horizons to signal their HAPOS to place their physical structures in a rest state and activate maintenance utilities. Narmer chose this user-friendly IT as the most effective for ensuring his Homo Sap groups' seamless adoption of his civic CalendarITIS. The ImageITIS prompts were part of their standard survival procedures.

Narmer superimposed the primary Sun cycle—the primary IT control for his civic CalendarITIS—onto the star Sirius cycle, the primary IT control of his survival CalendarITIS. He matched the period of the primary Sun cycle to that of the star Sirius' cycle of 365 light-and-dark periods of secondary Sun cycles or solar days, the Earthly period between the consecutive appearances of the Sun.

Though both versions cycled through numerically identical primary periods, the implementation of their applications depended on the appearance of different ImageITIS products in the Cosmos. The civic CalendarITIS depended on cosmic events: first, on the positioning of the Sun and Earth relative to each other in the primary Sun cycle, and then on the beginnings and endings of the light periods of a solar day. The survival CalendarITIS depended on both cosmic and Earthly events: The appearance and disappearance of the star Sirius from the eastern horizon of Ancient Egypt, together with the environmental conditions that the EOS established along the Nile River.

The differences allowed Narmer to implement the control applications of his civic CalendarITIS. To ensure predictability, he artificially fixed the IT of the primary Sun cycle at exactly 365 light and dark periods, or Solar days, defining secondary Sun cycles. His original analysis of the cosmic cycles had determined that the real period of the primary Sun cycle was about one quarter of a Solar day longer than 365. The outcome of ignoring the quarter Solar day was the gradual misalignment of the survival and civic CalendarITIS. The misalignment was critical to the effectiveness of the Thought-processing controls he designed into the applications of the civic CalendarITIS.

The applications of the survival CalendarITIS focused on food production. Their implementation depended on the outcomes of the EOS utilities controlling Earthly conditions along the Nile River. The ImageITIS products functioned only to alert his food production cadre to execute specific production procedures. The three minor cycles, though set at 120 Solar days each, could begin and end a few Solar days more or less than the 120. The periods were primarily guidelines. Changing the ITs or the applications would impact HAP survival, automatically triggering the HAP survival imperative and forcing his Homo Saps to depose him. All he could do was supervise the cadre and ensure that they responded to the repeating ImageITIS products with the correct applications

The design of the applications of his civic CalendarITIS allowed him to execute them on any day he chose. The applications focused on controlling the activities and Thought processing of his Homo Saps without immediate connection with HAP survival. Only he or his representatives delivered the directives that prompted his Homo Saps to execute the applications. Only he could change the applications in any way and could associate the outcomes with HAP survival or any other objective he chose. The operation of the civic CalendarITIS depended only on his count of the number of Solar days that had already passed since the beginning of each cycle of 365 Solar days.

He had to make similar adjustments to the real cycle fixing the cosmic position of the Moon relative to Earth. His analysis of the moon cycle described the period as a portion of a Solar day more than the 29 Solar days between the beginning of one Moon cycle and the beginning of the next. He artificially fixed the Moon cycle IT at exactly 30 Solar days and specified the primary Sun cycle of his civic CalendarITIS to contain 12 Moon cycles plus five separate Solar days.

He made his civic CalendarITIS more user-friendly for his Homo Sap groups to adopt with storytelling products he designed with the ITIS tool of OralITIS. He embedded the storytelling products as IS components in the Thought structures specifying the artificial IT he had fixed for his civic CalendarITIS. The storytelling products contained procedures and activities that he directed his Homo Saps to complete on any specific Solar day or period of Solar days.

As the only member of his Homo Sap group completely free to sequence and process any Thought structures in his individual Homo Sap memory, he functioned as the principal storyteller of the Homo Sap groups. As storyteller, as well as Homo Sap leader, he sequenced and disseminated Thought structures whose outcomes relieved individual Homo Saps of the constant pressure of their Homo Sap Thought-processing imperatives to sequence and process Thought structures other than those defining their immediate HAP survival procedures. The distraction from their HAP survival procedures left individuals as well as groups exposed to the dangers of predator attacks and other environmental hazards.

The Thought structures and IS components he embedded in his storytelling products using OralITIS protocols were the only Thought structures other than those defining HAP survival procedures that his Homo Saps processed. They trusted him to disseminate only Thought structures whose procedures and IS components kept them safe. The successes of his survival CalendarITIS reinforced that trust.

Some storytelling products focused on directing his Homo Saps to complete activities and functions he considered necessary for the government of the safe havens. He assigned cadres the responsibility for completing the activities and functions and initialized a civil service that would help him supervise the development and orderly operation of the safe havens. The new cadres followed his directives the same as did the cadres he assigned to operate his survival food production system. He regimented them to form new institutions dedicated to processing Thought structures whose procedures and protocols he designed to accomplish directed outcomes. They implemented applications for controlling the distribution of survival food resources, the collection of taxes, and other essential functions. The IS components always emphasized his godly authority to regiment his Homo Saps and direct them to undertake any activity he chose for them.

The most effective Thought-processing controls were those in the storytelling Thought structures that he developed with outcomes identifying extraterrestrial entities with special relationships with Homo Saps. Storytellers labelled them as "gods." The IS components contained a plethora of messages, many embedded in each other. All ultimately described the gods as all-powerful entities who operated in an environment free of the physical limitations of the Earthly environment and who controlled the survival conditions of the Earthly environments in which they confined Homo Saps and all other animal species.

One set of messages attributed the initiation and control of all cosmic and Earthly events to the arbitrary actions of the gods. Others described Homo Saps as "godly," the offspring of the gods and Narmer a god-king, the offspring of the chief god and the gods' Earthly representative. The gods assured Homo Saps of their survival as long as they obeyed the godly directives that their leaders disseminated. Homo Saps had the power to influence the gods by acceding to godly directives that Narmer claimed the gods transmitted to his individual Homo Sap memory. The gods directed him to disseminate the godly Thought structures and messages using storytelling procedures and protocols.

Narmer identified the IT of his civic CalendarITIS with the gods. The messages described the controlling cycles of the Sun and the Moon as the outcomes of godly actions. He and his predecessor storytellers had first identified the gods only in storytelling Thought structures. Homo Saps processed them in their individual memories during the dark periods when they switched their HAP survival system to maintenance and repair mode. The gods were without reality during the light period of the Solar day. Narmer gave them a reality with his designation of the Sun, Moon, and stars, the controllers of his civic CalendarITIS, as godly domains. These cosmic components functioned as ImageITIS products alerting Homo Saps to the continuous godly—and Narmer's—supervision of their activities in the real Earthly environment of a light or dark period.

Narmer designed the godly ImageITIS products to function as memory prompts for retrieving the Thought structures describing each god's attributes. Without Earthly godly realities to restrict him, he expanded the godly IS components with messages describing new Thought-processing directions.

He chose seven gods as the primary operators of the IT of his civic CalendarITIS and described them as his godly directors in designing the Thought-processing controls. Only these seven gods—through him as their Earthly representative—had the authority to make any changes to the IT or any other aspect of his civic CalendarITIS. Homo Saps named the seven gods he chose as Ra, Thoth, Osiris, Isis, Seth, Nephthys, and Horus, all siblings, all godly outcomes of the unbridled sexual activities of their godly father and mother, the cosmic gods Geb and Nut.

The seven operated as a hierarchical godly team. Each had a specific set of attributes whose outcomes—when they chose to execute the procedures expressing the attributes—helped or hindered Homo Saps in their survival activities. Narmer identified each god with Earthly ImageITIS products specifying the outcomes of their individual attributes. The IS component of the ImageITIS product describing a Solar day with or without an overcast of clouds obscuring the Sun attributed the event to a god's reaction to an event Homo Saps had initiated or failed to initiate. The IS component of the ImageITIS product describing a disaster occurring in one safe haven attributed the event to a god's displeasure with the Homo Saps who operated in the safe haven. The IS component in the ImageITIS product describing the relief from the disaster attributed the relief to the beneficence and mercy of another god intervening on behalf of the Homo Saps in the safe haven. Associated ImageITIS with individual gods allowed Narmer to give a physical reality to the god in the Earthly environment outside individual Homo Sap memories. The ImageITIS products changed the gods from insensible, ephemeral entities existing only in individual memories to sensible entities with a real Earthly presence.

Narmer associated the insensible cosmic god Ra with the sensible Earthly reality of the Sun. Storytellers labelled Ra "the Sun god." Narmer initialized the Sun as the IT component with a subset of cosmic operating utilities for controlling all physical events on the Earth and Moon and all other cosmic components of the Solar System.

He described the cosmic god Ra as the most potent of the seven godly siblings, with the exclusive godly attribute that allowed him to control the Sun's operating utilities. This godly attribute gave him the power to direct the Sun's execution of cosmic operating utilities to produce any Earthly physical conditions impacting the survival of Homo Saps and all other Earthly species.

Narmer associated the insensible cosmic god Thoth with the sensible Earthly reality of the Moon. The storytellers labelled him "the Moon god," but the cosmic god Ra alone controlled the amount of sunlight the Moon reflected to Earth to form the ImageITIS products determining the cyclical phases of the Moon and defining the twelve secondary cycles in the Sun's primary cycle.

Narmer associated the five youngest of the godly siblings, the insensible cosmic gods Osiris, Isis, Seth, Nephthys, and Horus, with the sensible Earthly reality of the five Solar days with which he completed the primary Sun cycle, the controlling IT of his civic CalendarITIS.

Narmer initialized his first set of Thought controls' ImageITIS products as the Sun, moon, and five Solar days at the end of the primary Sun cycle and their association with the seven gods. The appearance of the Sun to signal the beginning of the light period of a Solar day prompted his Homo Saps to retrieve and process the Thought structures he had sequenced to describe Ra's eternal control as Sun god of their Earthly survival. They all processed the same standard ImageITIS products wherever they operated along the Nile River. He modified the products to include IS components disseminating messages proclaiming his god-given suzerainty over all. The ImageITIS product of the Sun was a much more user-friendly and incorruptible carrier of messages than OralITIS products in his version of the ITIS tool that all Homo Saps had yet to make their own. The Sun's repeated appearance in the brightening Cosmos constantly broadcast the messages, constantly prompting them to process his Thought structures.

The appearance of the moon and stars prompted Narmer's Homo Saps to retrieve the Thought structures describing Thoth's godly functions. At the end of 360 Solar days of a primary Sun cycle, he transmitted the Thought structures describing the next five Solar days as those proclaiming Osiris, Isis, Seth, Nephthys, and Horus as Earthbound gods. Narmer gave every Solar day of his civic CalendarITIS immutable controls with the seven gods that only the gods through him could modify.

Narmer's god-king Thought processing directed him to expand the godly Thought structures. Giving names to the gods and identifying them in user-friendly ImageITIS products gave them a reality outside individual Homo Saps memories. He introduced Thought sequences describing the attributes of the seven gods and the Earthly events that identified the outcomes of the expression of their attributes. Storytellers labelled the sequences "epithets." Ra and Thoth had attributes mostly focused on their functions as cosmic gods and though the ImageITIS products of the Sun and Moon proclaimed their continuous presence and powers, they still existed primarily as Thought structures. The five Earthbound siblings had attributes more closely related to those of Homo Saps and operated with HAP restrictions to which Homo Saps could relate.

Storytellers designed ImageITIS products to identify and describe these gods. The IS component of the ImageITIS product captures the cosmic god Geb in hominid animal form. He lies on his back, covering a large portion of the surface of a globe representing the cosmic component of Earth. He faces Cosmos with the hominid animal sexual organ prominent and fully extended, rigidly erect, pointing to Cosmos. The cosmic god Nut hovers directly above him in Cosmos, facing him in the attitude of a female hominid animal ready to mount him and execute sexual procedures.

Capturing Nut and Geb, the godly mother and father of the five Earthbound gods, in large hominid animal forms dominating Earth, allowed the storytellers to include expanded IS components with subtle messages. One message made a direct connection between the gods and Homo Saps, establishing Homo Saps' godly ancestry.

Storytellers gave the four elder siblings, Osiris, Isis, Seth, and Nephthys godly attributes less real to Homo Saps. Their godly attributes allowed them to influence the survival conditions in the Earthly micro-environments in which Homo Saps operated. Homo Saps connected Osiris, Isis, Seth, and Nephthys with attributes that influenced water for plant growth, abundant harvests, the safe production of healthy offspring, and other events and survival conditions that depended on the uncertain outcomes or execution of the COS and EOS utilities controlling them.

Narmer focused his storytelling Thought structures more on the outcomes of the sexual and Thought-processing attributes of the Earthbound god Horus than on those of the four elder Earthbound siblings Osiris, Isis, Seth, and Nephthys. Only Horus operated in the Earthly environment without a compatible godly sexual partner. As with their cosmic godly parents, Osiris, Isis, Seth, and Nephthys operated as monogamous pairs.

Storytellers made the connection with Horus real. His highly productive sexual activities with many female hominid animals established his Homo Sap offspring as the first Earthly entities exhibiting ranges of sexual procedures and Thought-processing characteristics. The offspring he produced from his coupling directly with pure hominid females exhibited more powerful sexual and Thought-processing capabilities than those of offspring of pure hominid animals. His offspring had access to his Thought structures identifying themselves with more or less godly attributes but without godly capabilities for expressing any of them. They had access to the godly Thoughtbase but had to plead with their godly aunts and uncles to execute the godly procedures that improved HAP survival.

Storytellers who label themselves "astrologers" claim to be able to foretell the physical and Thought-processing attributes of individual Homo Saps from the dynamic state of the COS and EOS as they formed into viable adults.

Narmer fixed and established these storytelling Thought structures as another set of Thought-processing controls. His success with the survival CalendarITIS established his godlike Thought-processing capabilities. His freedom from immediate physical survival activities allowed him to execute his sexual procedures, as did the Earthbound god Horus, with many Homo Sap females and produce many offspring, With these demonstrations of his Horus inheritance, he reinforced his claim to god-kingliness, the direct offspring of the Earthbound god Horus coupling with his Homo Sap mother. With the improvements in their physical survival, his Homo Sap groups accepted his proclamation and the godly authority of his civic CalendarITIS.

He intentionally designed his civic CalendarITIS as an artificial Thought construct. The construct acted as an ITIS platform on which to hook and implement applications whose outcomes focused only on implementing Thought-processing controls instead of the practical applications of his survival CalendarITIS. He designed the first applications as directions for activities that Homo Saps would undertake on any Solar day he chose to solicit the help of the seven gods. Storytellers labelled the activities "praying."

The real primary cycles of the Sun and the star Sirius repeated in periods slightly different from each other. The COS extended both fractions of light-and-dark periods longer that the 365 light-and-dark periods that he fixed for the primary Sun cycle. He used their ultimate misalignment to achieve his Thought-processing control outcomes.

He chose to align the initializing implementation of his civic CalendarITIS with the survival CalendarITIS. His ITISanian analyses had determined that his Homo Saps captured and processed new Thought structures more readily when he sequenced the new Thought structures as extensions of the Thought structures they had already captured and automatically processed in their individual Homo Sap memories as part of their Earthly operations.

The applications of his survival CalendarITIS focused on organizing and regimenting the activities of a relatively small cadre of Homo Saps that he assigned to the tasks of producing and storing food survival resources for his growing group of Homo Saps operating in his safe havens. They captured the new ITIS Thought structures and processed them as extensions of the survival CalendarITIS. They had already acquired skills with this ITIS tool. Because of their trust in Narmer and his successful implementation of the survival CalendarITIS, they automatically processed the new Thought structures executing their procedures.

He expanded the applications of his civic CalendarITIS to focus on organizing and regimenting his Homo Saps to execute any activity he chose on any Solar day he chose. The division of the primary Sun cycle of the civic CalendarITIS into 12 secondary cycles with periods of a fixed number of 30 light and dark periods allowed him to design repeating events for specific light-and-dark periods in each secondary cycle.

He disseminated Thought structures whose IS components directed all his Homo Saps to coordinate their activities and execute the procedures of the events on the same light and dark periods. He designed the first events as "celebrations." The procedures instructed his Homo Saps to focus their Thought processing on Thought structures whose IS components praised the seven gods and, more specifically, himself as their god-king, for their successful survival. The IS messages described the seven gods and himself as their Earthly godly representative as the omnipotent, omniscient controllers of their physical survival. During those events he effectively controlled their group Thought processing and forced them to operate as one group to achieve the same outcomes, the cohesion and stability of his Ancient Egyptian Homo Saps civilization and their acknowledgment of his absolute authority as god-king.

He appointed his offspring as his representatives to implement the organizational and regimentation applications of his civic CalendarITIS in all his growing safe havens. They supervised and directed the activities of Homo Saps who operated outside the effective range of his hominid animal embedded oral technology. He specified their task as the dissemination of exact replicas of his OralITIS products. The fixed cycles of his civic CalendarITIS specified the fixed order with which they directed Homo Saps in Earthly locations far from his immediate supervision and physical presence. The OralITIS products his offspring disseminated contained IS components describing them as having his godly authority to enforce the outcomes of his directives.

Each OralITIS product contained the same messages proclaiming his status as the absolute god-king. He directed his offspring to function as his surrogate priests who conducted the fixed rituals of the godly celebrations and as administrators who organized and supervised the distribution of food survival resources.

As with the enhanced productivity outcome of his group's coordinated execution of the same food resource production procedures, his Homo Saps' processing of the same Thought structures with which he specified the organization of his safe havens enhanced the operational order wherever they operated. All his Homo Saps executed interlocking procedures ensuring individual safety but whose outcomes also ensured stable, predictable operations in the safe havens and the survival of the group as an interactive collective of individual Homo Saps. They all processed similar Thought structures with the same messages and outcomes.

The growing numbers of his Homo Saps made the maintenance of order in the safe havens more complex. To assist his supervising offspring, he organized cadres of Homo Saps to execute fixed sets of procedures focused on outcomes necessary to ensure compliance of the larger group with requirements of a stable social order. With his offspring as the immediate supervisors of the cadres, he formed them into the initializing kernel of social institutions dedicated to social order outcomes. He established the first priestly and military institutions, both as subsidiaries of the kingly/aristocratic institution, all of whose members were his offspring.

He set the primary function of the members of the priestly institution as that of his official storytellers. The Thought structures he directed them to process specified the procedures and protocols for capturing and disseminating fixed sets of messages that he authorized. The protocols gave the priestly Homo Saps the freedom to sequence any Thought structures but specified the messages that the storytelling Thought structures disseminated. The protocols ensured all messages proclaimed the omniscience and omnipotence of the seven Earthbound gods as Homo Saps' ultimate protectors and his absolute authority as their Earthly representative for disseminating godly Thought structures in the Earthly environment. As their god-king, any Thought structure he chose to sequence and directs them to process was a godly Thought structure.

He assigned the cadre of Homo Saps who formed his military institution the responsibility for ensuring the physical security of the safe havens and for maintaining the social order in accordance with protocols he specified. The protocols directed his military cadre to confine their Thought processing to a single Thoughtbase that specified sets of procedures. The first directed them to repel invasions of other Homo Sap groups who threatened the safety or survival resources of his safe havens. The second directed them to suppress any of his Homo Saps who challenged his godly authority and disrupted orderly operations in his safe havens.

The fixed order of his civic CalendarITIS provided him with the ITIS platform on which to organize and regiment the group activities of the religious and military institutions. The ITIS applications he designed specified the procedures for rituals that members of each institution completed during specified periods in the specified cycles of his civic CalendarITIS. The members of each institution displayed ImageITIS products and disseminated OralITIS products whose embedded Thought structures reinforced the messages describing the survival of Homo Saps as dependent on the efficient operation of the two institutions.

All messages triggered Homo Saps' fear for their individual survival and simultaneously allayed their fears with promises guaranteeing their survival, provided they complied with the godly directives of his religious institution and the physical safety directives of his military institution. The ultimate protocol he designed prohibited all future god-kings from making any changes to the cyclical structures and operation of his civic CalendarITIS as the primary organizational and regimentation tool for stabilizing the operation of his safe havens.

His descendant god-kings maintained the original IT specifications of the civic CalendarITIS without deviations for 50 Homo Sap generations after he implemented his civic CalendarITIS. They complied with the protocols that he set to prohibit them from making any fundamental changes to the order or periods of the anchoring artificial cycles.

The protocols directed them to ignore the differences between the actual periods that the COS set for the Sun, the moon, and the star Sirius cycles, and the fixed cycles he set for the primary and secondary cycles of his civic CalendarITIS. Narmer's imposition of the artificial IT of the fixed cycles as Thought-processing controls for regimenting the activities of the Homo Sap groups, together with their designation of the priestly and military cadres as their representatives for disseminating their control Thought structures and enforcing their procedures and outcomes, compelled Homo Saps who operated in the safe havens to comply with the regimentation features.

Homo Saps' compliance marked a fundamental change in their operating protocols. Instead of conforming their Earthly operations to the restrictions of the COS and EOS, they began to free themselves of the restrictions, executing procedures whose outcomes indicated their capabilities as Thought processors to ignore or overcome the restrictions and enhance their physical survival. They identified the civic CalendarITIS as a survival tool, but one whose survival successes depended on the outcomes of Thought processing rather than—and separate from—the uncertain outcomes of the COS and EOS.

6:

ANCIENT EGYPTIAN ITIS TOOLS OF WRITINGITIS AND AlphabetITIS

Dear Jim:

The growth in the numbers of Homo Saps who chose to operate in Narmer's safe havens and the great improvements in the survival rate and security of the safe havens that they attributed to his development and implementation of the organizational and regimentation applications of the civic CalendarITIS, confirmed one of the outcomes of his original ITISanian Thought processing: They demonstrated the overwhelming stabilizing power of disciplined, collective actions of groups of Homo Saps who coordinated their Thought processing to simultaneously execute the same procedures to produce the same outcomes. The stabilizing power was the outcome of his imposition of rigid specifications on the design, implementation, and operation of his CalendarITIS tools.

His ITISanian Thought processing began to identify and analyze the features of the last two ITIS tools, WritingITIS and AlphabetITIS. To reinforce the stability of his civilization, his Homo Sap leadership and god-king Thought structures directed him to implement rigid controls on the designs, applications, and implementations of all five ITIS tools.

He and his predecessors processed Thought structures they retrieved from two Thoughtbases to which only they had free access: one godly, the other ITISanian. The protocols from the godly Thoughtbase directed them to rigidly enforce the organizational and regimentation applications of the civic CalendarITIS and to destroy the HAP structures of any Homo Saps who processed Thought structures with procedures and outcomes corrupting those of the applications of the civic CalendarITIS.

Most of the Homo Sap leaders who led the Homo Saps for 50 generations after the implementation of the civic CalendarITIS operated exclusively as godly authoritarians. They ruthlessly eliminated Homo Saps who used the freedom and physical security of the safe havens to sequence and process in their individual Homo Sap memories Thought structures different from those that Narmer had authorized and disseminated.

The constant threats to the viable operation of their HAP operating systems forced Homo Saps to comply with the primary demands of their HAP survival imperative. They submitted to their god-kings' Thought-processing controls. The priestly and military institutions monitored and enforced their compliance. Their HAP survival conditioning their Thought processing, individual Homo Saps sequenced Thought structures with IS components whose messages described the benefits of operating in the physical security and survival resource riches of the safe havens as far outweighing the intangible benefits of freely processing Thought structures of their own individual design in their individual Homo Sap memories. They ignored the demands of their individual Thought-processing imperative, suppressing any Thought structures that their god-kings' institutions identified as subversive and detrimental to operating system stability.

Most Homo Saps preferred to adopt this submissive behavior in return for HAP survival throughout their development. They followed their institutions and resisted change. But they were unable to control unanticipated changes, the outcomes of the continued rivalry between the cosmic gods Thoth and Ra for the control of Homo Saps' Thought processing.

NARMER DEVELOPS THE ANCIENT EGYPTIAN VERSION OF WRITINGITIS

The outcomes of the cosmic god Thoth's actions provided storytellers with the first evidence of his rivalry with his elder brother Ra. Thoth had been monitoring Homo Sap development ever since his five siblings chose to confine themselves to the Earth environment. Narmer's establishment of the Ancient Egyptian civilization contained indications that Homo Saps had reached a Thought-processing stage that allowed them to begin weaning themselves of godly submission. Most Homo Saps exhibited some godly traits, the outcomes of the prolific sexual activities of his five siblings with hominid animals.

His first action was to transmit a set of Thought structures to Narmer. The Thought structures contained IS components with the messages simply stating that he, the cosmic god Thoth, had invented the ITIS tool of WritingITIS. Narmer, operating in his god-king Thoughtbase where he first received all godly messages, immediately processed the Thought structures, analyzing their compatibility with the god-king control focus. The outcome of the analysis left him confused. He responded with the Thought structure containing an IS component with a message that identified a danger with the implementation of the ITIS tool of WritingITIS.

Narmer described the danger in the development of WritingITIS as the loss of god-king control of Homo Saps Thought processing. Homo Saps would cease to rely on their Homo Sap memory to retrieve god-king-authorized Thought structures directing their Earthly activities. He delivered his directives in OralITIS products that Homo Saps captured, stored in their individual Homo Sap memories, and implemented at his command. When they retrieved his OralITIS products, they also retrieved the ImageITIS product describing him delivering the directives directly to them. Capturing his directives from WritingITIS products eliminated the impact of his physical presence. Homo Saps would tend to attribute the WritingITIS products as the primary sources of the directives, discounting his direct involvement as the source of the directives.

In addition, the acquisition of WritingITIS skills posed another control problem. Individual Homo Saps could capture their own Thought structures in WritingITIS products, making them appear to be those he authorized. WritingITIS skills would allow individual Homo Saps to make themselves independent Thought processors and direct themselves in their Earthly activities.

Narmer was uncertain on how to proceed with the development of the new ITIS tool. He awaited further direction from Thoth.

With the great expansion in both his Thought-processing facility and Homo Sap memory as the outcome of his analyses and development of the ITIS tools of OralITIS, ImageITIS, and the two versions of CalendarITIS, Narmer had made himself into the consummate ITISan. As principal Thought processor of his Homo Sap group, Narmer had developed the capability to multitask in the free space of his Homo Sap memory, processing Thought structures simultaneously as Homo Sap leader, god-king, and individual Homo Sap.

He had isolated his Homo Sap leader procedures and his god-king procedures in separate Thoughtbases that he stored in his individual Homo Sap memory. He sequenced the Thought structure describing the cosmic god Thoth's invention of WritingITIS to conform to the focus of each Thoughtbase. He implemented his ITISanian multitasking capabilities to process the Thought structure independently as Homo Sap leader and as god-king, allowing his responsibilities as Homo Sap leader and as god-king to separate the outcomes of his dual analyses.

The Thought structure he transmitted to the cosmic god Thoth described one outcome from his god-king analysis. Skills in the applications of WritingITIS would allow individual Homo Saps to capture their own Thought structures without his conditioning IS components, develop independent Thought-processing capabilities, expand individual Homo Sap memories, and free themselves of the Thought-processing controls of god-kings and their surrogate institutions.

Thoth declined to respond to Narmer's implied request seeking further direction on how he should proceed with the new ITIS tool. Thoth's silence sent a powerful message to Narmer: He was free to develop WritingITIS in any way he chose, a momentous event for any Homo Sap.

Thoth's omission of a definitive godly response was intentional. He prohibited himself for his own godly survival from using legitimate Atumic capabilities to influence Earthly events directly. His only safe option was to influence Homo Sap Thought processing. His analysis of the WritingITIS Thought structures he retrieved from MEMORY indicated that the outcome that Narmer as god-king produced was the most likely from all the possible outcomes.

Simultaneously, his analysis of the improving Thought-processing capabilities of his elder godly brother the cosmic god Ra, the supreme god of the Solar System and his supervisor, indicated that Ra would reach the same outcome when he monitored, though infrequently, events on Earth. Ra would identify Homo Saps' acquisition of WritingITIS skills as a further threat to his absolute control of Solar System events. He would immediately ban the development of the new ITIS tool, stimulating Homo Saps' fear of his retribution before they expanded their Thought-processing capabilities and sequenced the Thought structures with IS components disseminating messages describing his godly threats as having no substance.

To annihilate them he would have to destroy Earth, which would destabilize the Solar System, which would cause extreme disturbances in the equilibrium of the Cosmos. These disturbances would give the supreme cosmic god Atum a cause to banish him into cosmic limbo.

Thoth kept monitoring Ra's actions.

Narmer accessed the infinite Thought store of MEMORY while he waited for Thoth's response and began multitasking. He first expanded his analysis of the ITIS tool of WritingITIS in the Thoughtbase defining his Homo Sap leadership responsibilities. His analyses produced outcomes similar to those he identified from his analyses in his god-king Thoughtbase but with extensions of the IS components. The extensions contained messages identifying the costs and benefits to his Homo Sap groups of the development of WritingITIS.

One outcome captured his Homo Sap leadership focus: The extensions disseminated messages that identified great benefits from the development of WritingITIS but also periods of killing chaos before the achievement of those benefits. The IS components connected his Thought processing to an outcome of his development of the survival CalendarITIS. The stimulation of Narmer's continued Thought processing was the outcome that Thoth had intended with his silence in responding to Narmer's request for further direction.

The benefits were similar to those his Homo Saps had independently developed after his implementation of the survival CalendarITIS. He had assigned the responsibility for executing survival procedures to cadres of Homo Saps who dedicated themselves to the production of food survival resources. Over many Homo Sap generations, the dedicated cadres had made great improvements in the productivity of the food resource production procedures. Individual Homo Sap members of the cadres had freely processed Thought structures describing their production procedures, developed and implemented most of the improved procedures, demonstrating the powerful outcomes of the Homo Saps' Thought-processing imperative.

Their processing of the food production procedures in their individual Homo Sap memories had expanded their Homo Sap memory capacity and Thought-processing capabilities, giving them the free space to re-sequence the procedures. They made them more user-friendly for individual Homo Saps to execute with less energy but with improving productivity. The connection of this outcome to his analyses of the WritingITIS Thought structures prompted him to sequence an IS component with the message that the acquisition of WritingITIS skills by individual Homo Saps allowed individual Homo Saps to produce procedures whose outcomes enhanced Homo Sap survival.

Narmer developed an application of his multitasking ITISanian skills that allowed him to transfer Thought structures to and from the Thoughtbases he stored in his Homo Sap memory, meld them in the free space, and process them. He executed the application with two sets of Thought structures.

One set, from his god-king Thoughtbase, defined the advantages of WritingITIS in helping him as god-king, to stabilize and establish order in his Ancient Egyptian civilization as well as control Thought processing in the growing numbers of Homo Saps who chose to operate in the safe havens. The other set of Thought structures, from his Homo Sap leadership Thoughtbase, described the freedoms that individual Homo Saps would obtain from his implementation of WritingITIS. The two were incompatible. He was unable to meld them to meet both his god-king requirements for stability and order and his Homo Sap leadership requirements for enhancing individual Homo Sap freedoms and survival.

His god-king outcomes directed him to implement the ITIS tool as a Thought-processing control and dissemination tool that allowed him to ensure that all his widely separated and growing Homo Sap groups followed his god-king directives precisely. Using OralITIS products to disseminate the precise directives left him open to the dangers of inadvertent changes in the Thought sequences and control procedures as one Homo Sap transmitted them to another, using their own variable OralITIS skills. Capturing his directives in WritingITIS products ensured that each of his surrogates supervising distant safe havens implemented the same unaltered procedures.

His Homo Sap leadership Thought structures directed him to make the WritingITIS tool as user-friendly for all his Homo Saps as was OralITIS. His ITISanian analysis produced the outcome that individual Homo Saps would automatically make WritingITIS as user-friendly and ubiquitous as OralITIS. This imperative to make ITIS tools as user-friendly to individual Homo Saps as possible was a characteristic that they would demonstrate at every stage of their development.

His processing of the combined Thought structures defining his Homo Sap leadership and god-king responsibilities indicated that his priority was to ensure the stability, orderly development, and longevity of his civilization. He had to ensure that all his Homo Saps obeyed his god-king directives whether captured in OralITIS or WritingITIS products. He imposed his first control by declaring that WritingITIS was a sacred godly tool whose skills only he and the surrogates he appointed had the godly authority to acquire and practice.

He began to establish the specifications of the Ancient Egyptian version of WritingITIS. Homo Saps had been scribing ImageITIS products on stone, wood, and other ITIS substrates for hundreds of Homo Sap generations before that of Narmer's. They designed the ImageITIS products to transmit messages to other Homo Saps, mostly about available survival resources or dangers to HAP structures. The ImageITIS products were often simple sketches. They outlined the shapes of animals and objects but without consistent designs of the IT and IS components of the products. One group of Homo Saps had difficulty capturing the messages in the ImageITIS products that other groups produced.

His ITISanian analyses indicated that, as communication devices for transmitting Thought structures, their primary deficiencies were the bugs in their IT. Few Homo Saps had the physical skills, the hand-eye coordinating skills, to reproduce exact copies of the same objects. The bugs manifested themselves as differences in the IT designs that individual Homo Saps introduced as they reproduced the ImageITIS products. To make them effective transmitting devices he needed to fix the IT designs and specify the IS component of each ImageITIS product.

Narmer began to standardize sets of ImageITIS products, fixing the IT and IS components. He established their function as memory prompts. He keyed them to Thought structures that he sequenced to define procedures and disseminate specific messages. Capturing his standardized ImageITIS products with their HAP visual technology would prompt Homo Saps to retrieve from their Homo Sap memories the IS components specifying the procedures and messages he associated with each ImageITIS product.

Homo Saps labelled the IT as "hieroglyphics" and the messages that the IS components of the Thought structures disseminated as "sacred."

The IT specifications he established allowed him to design hundreds of unique ImageITIS products. He compelled those Homo Saps he designated to acquire WritingITIS skills—storytellers labelled them "scribes"—to store the ImageITIS products and their associated procedures and messages in their individual Homo Sap memories. He designed protocols specifying the procedures with which scribes processed any sequence of ImageITIS products. The protocols specified how they were to capture the directives and messages he would have transmitted to them in OralITIS products. Reprocessing the directives and messages in their individual Homo Sap memories, they re-sequenced them to emulate the OralITIS products that he would have produced to deliver the directives and messages directly to them. The ImageITIS products in their sequences prompted them to retrieve Thought structures describing him delivering the directives and messages directly to them in OralITIS products.

The sequences of ImageITIS products reprocessed into OralITIS products in accordance with his protocols impacted them with the authority of his physical presence wherever they operated outside the range of his embedded HAP vocal technology. He could scribe sequences of ImageITIS products on ITIS substrates, give them to a Homo Sap acting as a carrier IT to deliver to his surrogates, and expect them to implement any procedure or to act on any message he embedded in the IS components.

Narmer now had four ITIS tools, OralITIS, ImageITIS, CalendarITIS and WritingITIS with which to implement Thought-processing controls in any safe haven in his civilization His ITISanian Thoughtbase contained all their specifications. Networks of Thought structures described the potential impacts of their features and applications on the effectiveness of his supervision of activities in his safe havens.

The primary message Narmer embedded in each ImageITIS product proclaimed the absolute authority of the god-king. The sequences contained in the secondary messages specified the directives and procedures that the god-kings wanted Homo Saps to implement. Narmer and his successors had to design thousands of ImageITIS products, each containing IS components focused on precise directives and procedures. The Homo Saps he designated had to devote at least 12 primary sun cycles of his civic CalendarITIS to capture and store the hundreds of ImageITIS products, their embedded messages, and the communication protocols in their individual Homo Sap memories to function as effective scribes.

WritingITIS was Narmer's most powerful remote-control ITIS tool. But his access to MEMORY had identified a fifth ITIS tool—AlphabetITIS—whose application was as potentially powerful for disrupting both his god-king controls and the stability of his growing civilization as WritingITIS.

NARMER SUPPRESSES THE DEVELOPMENT OF ALPHABETITIS

He identified the Ancient Egyptian AlphabetITIS as having only 24 unique ImageITIS products. He could sequence them into many different structures to capture and disseminate any procedures or messages he chose. They were much more user-friendly than the hundreds of ImageITIS products he had specified for Ancient Egyptian WritingITIS.

The design of the 24 ImageITIS products as WritingITIS elements left them free of any association with gods or god-kings. Their user-friendliness allowed individual Homo Saps to acquire them quickly without the Thought-processing conditioning of his priestly institution during the 12 sun-cycles. They gave individual Homo Saps the freedom to independently structure messages that differed from those that the god-kings authorized and to disseminate them in WritingITIS products for other Homo Saps to process.

Without the ITIS tool of AlphabetITIS, individual Homo Saps transferred Thought structures and messages they sequenced in OralITIS products only. His successor god-kings and their two institutions could easily suppress their dissemination with HAP killing procedures that destroyed the HAP structures of those Homo Saps who persisted in producing and disseminating OralITIS products with disrupting messages. But with the AlphabetITIS tool, individual Homo Saps could ensure the continued dissemination of their ungodly Thought structures and messages in WritingITIS products after the god-kings and their institutions executed killing procedures to dissemble their HAP structures or banned them from operating in the safe havens.

Narmer, focused only on the stability and longevity of his civilisation, chose to forbid the further development of his prototype AlphabetITIS and rigidly enforced the established specifications of his Ancient Egyptian version of WritingITIS.

He sequenced a set of Thought structures that directed his successor god-kings to prohibit the development of AlphabetITIS. The IS components described applications of AlphabetITIS as most dangerous to the god-kings' absolute control of Thought processing in his Homo Sap groups and to the orderly development, stability, and longevity of his civilization.

7:

Individual Homo Saps with WritingITIS Skills begin to Undermine Ancient Egyptian Institutional Controls

Most of Narmer's successors chose to focus their Thought processing on their functions as god-kings. They processed only the procedures of their god-king and subsidiary institutional Thoughtbases. They maintained and updated the operating system Narmer had initialized to enhance the coherence of the Ancient Egyptian civilization. They enforced the organization and regimentation applications of the civic CalendarITIS, grew their institutions, and maintained rigid control of the specifications and applications of the ITIS tools of OralITIS, ImageITIS, and WritingITIS. They led their disciplined military institution in hominid animal killing procedures to expand their safe havens to accommodate the growing numbers of their Homo Sap groups and to secure the safe havens from marauding Homo Saps of other groups who focused on pillaging the safe havens for survival resources. They processed primarily Thought structures whose procedures and outcomes ensured the HAP domination of their own Homo Sap groups over other Homo Sap groups that competed with them for the food survival resources of the Nile River.

They were highly successful in maintaining the growth and reinforcing the cohesion and stability of their Homo Saps civilization Ancient Egyptian Homo Saps developed over the first 50 generations of powerful Thought-processing controls with the applications of the civic CalendarITIS and of ImageITIS. Their successes trapped their Thought processing in a feed-back loop recycling through the same sets of godly and institutional Thought structures with the same control outcomes, a fault embedded in all rigidly-controlled institutionalized systems.

The institutionalized leaders were unable to break out of the Thought-processing loop and sequence new Thought structures whose IS components described potential disruptions in the orderly operation of their Homo Sap civilization. They failed to update institutional operating systems with routines that safely handled the outcomes of the Thought-processing imperative stimulating independent Thought processing in individual Homo Saps. Most disruptions followed the growth in the number of individual Homo Saps outside institutions who independently acquired ITIS skills, particularly those of WritingITIS.

Only a few of Narmer's successors chose to operate with equal focus on their group responsibilities as Homo Sap leaders as well as absolute god-kings. These few demonstrated powerful Thought-processing capabilities and chose to operate with equal effectiveness as Homo Sap leaders. They accessed their ITISanian Thoughtbase, processing and sequencing Thought structures whose procedures also ensured the HAP survival of the Homo Saps who operated in their expanding safe havens but without the drastic killing outcomes of the military procedures of their institutional Thoughtbase. They implemented and expanded the applications of the ITIS tools of OralITIS, ImageITIS, CalendarITIS, and WritingITIS whose specifications Narmer had fixed.

The ITISanian Thoughtbase that each god-king stored in his individual Homo Sap memory and freely accessed as principal Thought processors of the Ancient Egyptian civilization contained Thought structures describing ImageITIS products as the most powerful disseminators of Thought-processing controls to Homo Saps who practiced only the ITIS skills of OralITIS. These Homo Saps—the majority—captured Thought structures with their HAP aural and visual technologies. Both are embedded HAP survival technologies. Their primary programming focuses on filtering all OralITIS and ImageITIS products through survival interfaces for Thought structures whose IS components alert individual Homo Saps to imminent dangers.

Storytellers label such Homo Saps as "illiterate." They formed the vast majority of Ancient Egyptian Homo Saps, and the god-kings and their religious institution kept them illiterate. OralITIS and ImageITIS are powerful ITIS tools for disseminating Thought structures that trigger fear in Homo Saps. They tolerated this restriction on their Thought processing, relying on their leaders, as they had in the past, to supply them with survival resources or alert them to dangers.

The god-kings focused the design of the ImageITIS products on IT features that captured one set of unchangeable Thought structures in which they embedded IS components eternally disseminating one unvarying message: the absolute power and authority of the god-kings as the Earthly godly representatives of the cosmic gods over all Ancient Egyptian Homo Saps. All future institutions—particularly those of government and the Christian religious institution—implemented this powerful application of ImageITIS products to control Thought processing.

The Ancient Egyptian institutions chose the IT features of Earthly physical size and Earthly physical longevity with which to design the ImageITIS products. They designed them as massive monuments. Their physical size alone captured and disseminated Thought structures with fixed IS components disseminating messages. One set proclaimed the godlike power of the god-kings to modify and overcome the EOS utilities controlling and restricting the HAP capabilities of individuals.

The god-kings formed the ImageITIS products with stone, an Earthly material whose longevity under the inexorable recycling procedures of the EOS far exceeded that of any individual Homo Sap or Homo Sap generations. The massive stone ImageITIS products, unchanging and overwhelming puny HAP structures, automatically prompted Homo Saps to process in their individual Homo Sap memories Thought structures describing their HAP awe of Earthly events that they attributed to the actions of godly entities. The ImageITIS products disseminated the same Thought structures with the same messages to generation after generation, the messages eternal. They awed generation after generation of Homo Saps, compelling them to obey their god-kings, instantly and without equivocation, in fear of the Earthly power of the god-kings.

They developed ancillary ITIS tools that allowed them to permanently inscribe in the stone the hieroglyphic elements of the Ancient Egyptian version of WritingITIS. They captured in the hieroglyphic WritingITIS products Thought structures whose IS components repeated the godly messages embedded in the size of the ImageITIS products. Homo Saps were unable to avoid them. The awesome massiveness and permanency of the stone ImageITIS products functioned as memory prompts forcing them to continuously retrieve and process IS components whose messages repeated the god-kings claims to absolute godly authority over all Ancient Egyptian Homo Saps.

Hundreds of hieroglyphic elements, each a unique ImageITIS product, captured and disseminated fundamental Thought structures containing IS components whose fixed messages claimed the intimate connections of the god-kings with their seven primary gods. They sequenced the hieroglyphic elements into more complex Thought structures. They embedded IS components within IS components that disseminated messages within messages. They introduced sequences with subtle variations reasserting their claim to godly omnipotence and omniscience as Earthbound surrogate godly entities.

The constant dissemination of the same messages succeeded. Homo Saps reflexively submitted themselves to the Thought-processing directives of the god-kings. They consistently provided them with abundant food survival resources and assured them of safe havens in which they operated without HAP fear. The god-kings' rigid applications—through their religious and military institutions—satisfied both HAP requirements. That they operated under tight Thought-processing restrictions was irrelevant. Their leaders had satisfied their social contract.

The success of their ITIS applications prompted the god-kings to reprocess Narmer's original warnings of the dangers of widespread ITIS skills. They restricted the numbers of Homo Saps who they authorized to acquire ITIS skills to those they needed to ensure the efficient operation of their food resource and safety applications. They limited the ITIS skills of the vast majority of their subject Homo Saps to those of OralITIS. They captured and stored these in their individual Homo Sap memories, and processed only Thought structures that the god-kings and their institutions disseminated in OralITIS and ImageITIS products.

Though the god-kings satisfied their primary survival needs, Homo Saps still reacted to their HAP survival imperative. The demands of the imperative compelled Homo Saps to exercise other HAP survival procedures. The abundance of food survival resources in the safe havens left them with more freedoms to exercise less-immediate survival procedures. Again, the one set of survival procedures they had the freedom to exercise more frequently were the HAP sexual procedures. They were user-friendly and outside the direct control of their god-kings and institutions. They began to produce offspring, rapidly increasing the numbers of individual Homo Saps who operated in the safe havens.

The increasing numbers forced the god-kings to expand their religious and military institutions in order to satisfy their leadership obligations for HAP safety and distribution of survival food resource and presented the god-kings with a Thought-processing control dilemma. They needed more Homo Saps to acquire WritingITIS skills to help them manage and control survival resources, but the more Homo Saps who acquired ITIS skills the greater the risk of the loss of Thought-processing controls.

The god-kings at first confined the acquisition of WritingITIS skills to a select few of their offspring who they designated as supervisors of the institutions and controllers of the safe havens. This strategy worked while their group numbers were small. The god-kings' exercise of their HAP sexual procedures produced the numbers of offspring they needed to help them control the operation of their civilization.

To increase the numbers they needed for effective governance, they developed a new strategy. They gathered and secluded large numbers of female Homo Saps that Storytellers labelled "concubines." All dedicated themselves to functioning as exclusive god-king sexual partners for incubating and delivering god-king offspring. But the energy requirements of their HAP operating systems restricted the frequencies at which they could execute their HAP sexual utilities and procedures. They were unable to reproduce the numbers of offspring they needed to satisfy the governance and control requirements of the rapidly growing numbers of individual Homo Saps operating in their safe havens.

They attempted to overcome their Earthly HAP limitations with a powerful ITIS procedure, one that more and more Homo Saps began to emulate in attempts to influence their, and their offsprings', HAP survival. The ITIS procedure allowed the god-kings to give themselves one of the names of an Earthbound cosmic god. Each god had many names, each name capturing the Thought structure specifying one of the many Earthly powers of the god. The Ancient Egyptian god-kings initiated this ITIS procedure, focusing on their seven primary gods. Future Homo Saps—particularly those who belonged to the Ancient Greek and Ancient Roman subsidiary civilizations—emulated the Ancient Egyptians but expanded the number of individual gods, each exhibiting one or more of the powers of the seven Ancient Egyptian gods.

Assuming the name of a god automatically gave an individual Homo Sap the specific powers of the god. After the Ancient Greek Homo Saps completed the development of the most power ITIS tool—AlphabetITIS—the gods began to forbid Homo Saps from uttering their real names and hid their names in encrypted ITIS products. The gods took this action as a retaliatory procedure. Homo Saps Thought-processing developments had relegated them to minor roles in Earthly developments. Homo Saps began to assume godly powers and began challenging the gods.

For sexual powers the god-kings gave themselves the name of the cosmic god Horus. He had exhibited a prodigious sexual productivity in coupling with hominid animal females to satisfy his unAtumic sexual demands on his advent into the Earth environment as a hominid animal male. The outcomes of his sexual activities initialized the Homo Saps species of hominid animals.

The god-kings were still unable to match the output of the growing numbers of their Homo Sap subjects despite their acquisition of hundreds of exclusive sexual partners.

Their ITISanian Thought processing directed them to another strategy. They gave their priestly institution the responsibility for recruiting individual Homo Saps and conditioning their Thought processing to function as disciplined institutional scribes. The priests adopted the practices of Homo Sap parents in preparing their offspring to function in the Earth environment. They recruited Homo Sap offspring, mostly those of nobles but also other Homo Saps, as WritingITIS apprentice scribes whose Thought-processing capabilities were still malleable.

They set the period for ensuring the apprentices stored the precise design of the scribal Thoughtbase in their individual Homo Saps memories to twelve primary sun cycles of the civic CalendarITIS.

Each new scribe had to capture hundreds of ImageITIS products that formed the elements of the hieroglyphic version of WritingITIS and reproduce them unchanged. The apprentices graduated as disciplined individuals whose priestly-induced Thought-processing conditioning confined them to processing the rigid protocols and procedures of hieroglyphic WritingITIS and the scribal Thoughtbase.

This Thought-processing conditioning designed them to function as mechanical WritingITIS output devices. Their Thought processing and WritingITIS skills were dedicated to one outcome: The reproduction of exact facsimiles of authorized hieroglyphic ImageITIS products with their specified IS components and messages.

The increasing numbers of scribes had their intended impact on the efficiencies of the Ancient Egyptian operating system: The strict IT specifications allowed the god-kings through their institutions to improve their governance of the safe havens and control Thought processing.

Their ITISanian analyses of the successes identified the rapid dissemination of the WritingITIS products as critical to their continuing success. They began to pressure the scribes to increase their productivity. The outcome was a critical ITIS event that began the erosion of god-king and institutional Thought-processing controls and absolute authority.

The scribes identified the causes of their productivity limitations as the complex designs of the hieroglyphic ImageITIS products that the god-kings and their institutionalized representatives specified and enforced as the authorized elements of WritingITIS. Though the admonitions of their priestly supervisors during their long scribal apprenticeship made them fearful of changing any hieroglyphic specifications, the productivity pressure allowed a few to justify the release of their Thought-processing imperative from the priestly restrictions. They began to analyze procedures that were both user-friendly and maintained the sacerdotal context of their WritingITIS products acceptable to their priestly supervisors

The new Thought processing automatically and seamlessly expanded the free space in their individual Homo Sap memories. They sequenced Thought structures that identified factors that limited their WritingITIS productivity. They identified one factor as the discrepancy between the speed at which they processed Thought structures in their individual Homo Sap memories and the speed at which they transcribed and captured the Thought structures in WritingITIS products. As their Thought-processing driver, the Homo Sap Thought-processing imperative operated at speeds much faster than the maximum speed at which their hominid animal operating systems activated and coordinated the HAP technologies that they had to use to capture Thought structures in WritingITIS products.

The reproduction of the hieroglyphic ImageITIS products on WritingITIS substrates required the scribes to acquire unique skills in coordinating their HAP technologies for reproducing exact designs. The uncertainties of the EOS impacting their HAP development during their formation into viable HAP individuals stabilized only a few scribes with individual HAPOS capable of executing fully-functional utilities for driving and coordinating hand-eye actions that reproduced exact facsimiles of ImageITIS products on WritingITIS substrates. HAPOS stabilized with different sets of utilities for coordinating their physical actions. Each individual's HAPOS contained unique sets that allowed them to execute HAP actions with more efficiency than the HAPOS of another individual. Storytellers labelled those individuals with access to utilities that allowed them to reproduce ImageITIS products easily and quickly "artists and artisans."

The productivity analyses produced Thought structures that described only one user-friendly option for improving their WritingITIS product outputs. They were unable to control the speed at which their Homo Saps Thought processing operated or to modify the HAPOS utilities to make them more efficient in driving their HAP technologies faster. The most user-friendly option was the simplification of the designs of the hieroglyphic ImageITIS products. They chose to reduce the number of physical actions required to reproduce a hieroglyphic ImageITIS element. They simplified the IT of their WritingITIS. They devised ImageITIS products that required only a few physical actions to reproduce on WritingITIS substrates. They designed a new scribal Thoughtbase containing Thought structures they independently sequenced to describe the association of the unique but simple ImageITIS products with the more complex hieroglyphic ImageITIS products.

The unique, simpler ImageITIS products functioned as memory prompts for retrieving Thought structures describing hieroglyphic ImageITIS products that in turn functioned as memory prompts for retrieving Thought structures describing the godly messages that the god-kings embedded in the hieroglyphic ImageITIS products. The speed of their Thought processing compensated for the added complexity of accessing more than one Thoughtbase to identify and retrieve godly messages. Storytellers labelled the unique, simpler ImageITIS products "hieratic" to identify their sacerdotal context.

The new hieratic version of their WritingITIS allowed them to continue their easy operating mode. To produce their WritingITIS products they still sat down and, without great physical stress on their HAP structures and operating utilities, reproduced hieroglyphic and hieratic ImageITIS elements on lightweight WritingITIS substrates that Homo Saps labelled "papyrus." Some scribes also engraved the ImageITIS elements in stone, giving the Thought structures a permanency and immutability that directed Thought processing for generation after generation.

The hieratic ITIS event identified an agent of change that all future institutions were unable to avoid or contain. All institutional OS contained an undeletable bug, a virus, a Trojan horse, whose inevitable outcome forced institutions to express their institutional impotence in controlling the power of the Homo Sap Thought-processing imperative operating in the privacy of individual Homo Sap memories. The institutions themselves invariably trigger the execution of the virus. Twelve primary sun-cycles of acquiring WritingITIS skills allowed the scribal Homo Saps, with institutional encouragement, to expand their individual memories. Expanded memories gave them the opportunity to free their Thought processing from institutional controls.

The development of hieratic WritingITIS was the first indication of Homo Saps' inherent compulsion to redesign ITIS tools and make them user-friendly for individual Homo Saps to acquire and implement free of institutional controls. The compulsion was one that Homo Saps exercised only after the Ancient Greek Homo Sap Aristotle died, some 150 generations of Homo Sap Thought processing and ITIS developments after those that Narmer initialized.

The Ancient Egyptian institutions maintained absolute control of their civilization. The set of institutional Thought structures that the scribes processed contained messages threatening them with godly retribution, HAP damage, or ejection from the scribal institution for allowing themselves to sequence, process, and disseminate Thought structures and messages other than authorized ones.

The institutional threats forced most scribes to filter their Thought processing through the interface of their HAP survival directives. Their fear of damage to their HAP structures forced them to process only authorized Thought structures. The threat of ejection connected their HAP survival with their membership in the scribal institution. They made protecting that membership and the institution their controlling survival activities. They had one more incentive compelling them to operate within the bounds of their authorized Thoughtbases.

Outside the scribal and other institutions, the god-kings directed most Homo Saps to function as physical tools—manual laborers—of the god-kings. They undertook the Earthly tasks whose procedures and outcomes stressed and damaged their HAP structures. The parents of the scribes and the scribal institution constantly disseminated the Thought structures describing the physical survival advantages of those Homo Saps who acquired WritingITIS skills. They made continued membership in the scribal institution a survival imperative.

Most of their fellow Homo Saps labored in the fields along the banks of the Nile River to produce survival food resources or hauled massive blocks of stone for the construction of the gigantic ImageITIS products of pyramids, funerary complexes, and statues, suffering great physical stresses to their HAP structures in the service of their god-kings.

The minimal HAP stresses on the Homo Sap scribes while executing their WritingITIS procedures prompted other Homo Saps to sequence and disseminate Thought structures whose IS components described the scribes as privileged Homo Saps separate from the god-kings' offspring, most of whom functioned as members of the aristocratic, priestly, and military institutions. The IS component also contained the message that all and any Homo Saps had the inherent manual and Homo Sap memory capabilities to acquire WritingITIS skills.

HIERATIC WRITINGITIS INITIATES CHANGE IN ANCIENT EGYPT

The outcomes of the implementation of the hieratic IT for a second version of Ancient Egyptian WritingITIS demonstrated the speed at which Homo Saps adopt an ITIS tool. The hieratic ImageITIS elements were much simpler and easier to reproduce than their hieroglyphic sources. The relatively simple configurations allowed individual Homo Saps with minimal HAP hand-eye coordination skills to reproduce the hieratic ImageITIS elements on WritingITIS substrates.

These individual Homo Saps, most offspring of members of the aristocratic institution, began to produce hieratic WritingITIS products to capture Thought structures they sequenced in the free space of their individual Homo Sap memories. The ITIS process automatically expanded their Thought-processing capabilities with minimal conditioning from any association with god-king or priestly messages. The primary outcome, one that always followed an increase in the user-friendliness of ITIS tools, impacted hierarchical order and institutional controls.

The new scribes, free of scribal institutional controls, began to re-sequence god-king, priestly, and scribal institutional Thought structures. They began to produce new WritingITIS products containing Thought structures whose IS components disseminated messages challenging the absolute authority of the god-kings and their priestly surrogates. Disseminating these challenging Thought structures in WritingITIS products initiated a process that began to disrupt the orderly operation of the Ancient Egyptian civilization.

Most of the disruptions were the outcomes of the actions of new leaders replacing old leaders and new god-king replacing old god-king, sometimes forcibly and sometimes peacefully. Many of the new leaders and god-kings claimed their direct blood-line as offspring of previous god-kings to justify their assumption of supreme controls.

Some modified or eliminated operating procedures that predecessors introduced but left unmodified the core operating system maintaining stability.

They continued to use the primary institutional and ITIS tools, always maintaining the sacerdotal IS components and messages in the ITIS products. Some had developed their ITISanian analytical skills and determined that many of the disrupting changes were the outcomes the rigid institutional suppression of individual Thought-processing freedoms. They defused the disrupting potential of these changes by giving individuals more freedom to independently implement ITIS applications but under their supervision, ensuring that IS components disseminated messages benign to the god-kings.

Their ITISanian analyses also determined that the causes of some disruptions were the actions of the offspring they produced with the hundreds of female Homo Saps they had sequestered for their exclusive sexual activities. They established a new sexual reproduction strategy in order to identify the offspring who they could name as their legitimate successors.

They began to couple with their sisters, justifying the strategy as emulating the exclusive godly brother-sister pairing of the two Earthbound gods, Osiris and Isis. They claimed that the legitimate god-king offspring were those their sisters carried after a male god assumed a HAP structure to couple with her.

The offspring they produced with other female Homo Saps formed the aristocratic institution. All, though without complete god-king qualifications but as offspring of god-kings, claimed freedom from menial survival responsibilities. Storytellers labelled them as "nobles." Many acquired skills in the applications of the hieratic version of WritingITIS but without the Thought-processing conditioning of priestly or scribal institutions. These offspring, freely expanding their Homo Sap memories and Thought-processing facilities as they practiced hieratic WritingITIS skills, began to associate the simple hieratic ImageITIS elements with the sounds they produced to produce OralITIS products.

They began to produce WritingITIS products containing Thought structures that contained IS components disseminating secular messages using the user-friendly protocols of Ancient Egyptian OralITIS products. They began to assume the role of storytellers and produced WritingITIS products that described the events that disrupted the last of the fifty generations that followed that of Narmer. For most of those generations the god-kings and their institutions exercised absolute control of Thought processing in Ancient Egypt, but the last failed to meet their Homo Sap leadership obligations..

The last god-kings forgot that their tenure as god-kings depended on their Homo Sap subjects' satisfaction with their leadership obligations that guaranteed their HAP survival requirements. The new storytellers attributed the ensuing disruptions to "leadership failures."

The prosperity and stability, the tight institutional Thought-processing controls, and the successes of the regimentation features of the civic and survival versions of CalendarITIS gave few opportunities for the later god-kings to process leadership Thought structures and update the Homo Sap leadership Thoughtbase. They had allowed themselves to become complacent.

They failed to monitor Earthly environmental conditions in the safe havens for changes that the uncertainties of the EOS produced. They needed to develop new procedures that helped them offset the impact on the production of food survival resources. Instead, they relied on their priestly institution and their growing numbers of bureaucrats and scribes who, without direction from their god-kings, simply implemented and executed older, increasingly inefficient procedures that failed to optimize production in a deteriorating Earthly environment along the Nile River.

They operated only in the god-king Thoughtbase. They focused only on ensuring the safe havens they established along the banks of the Nile River remained loyal to them. They assigned mostly their offspring to supervise and enforce order in the safe havens. Storytellers labelled the safe havens "nomes" and the supervising Homo Saps "nomarchs."

They began to rely almost exclusively on WritingITIS products to disseminate directives to their subject Homo Saps. They assigned the responsibility for compiling the WritingITIS products and determining the directives, including critical IS components, to their priestly and bureaucratic institutions. The institutions had grown—as most institutions inevitably do—into moribund organizations, bloated with officials and scribes, few directly concerned with the delivery of institutional services. They compiled the directives into operating manuals specifying the procedures that the nomarchs and their governing institutions in the safe havens should execute in response to any event. They focused the directives and procedures more on sustaining the operations of the institutions than on improving the services for which Homo Saps originally allowed them to form. The fears that the first god-king Narmer expressed when the cosmic god Thoth announced his invention of WritingITIS were about to produce disastrous outcomes.

More and more individual Homo Saps had acquired WritingITIS skills as they now had to rely on the Thought structures in the manuals for direction. As Narmer had predicted, the manuals failed to emulate the authority of OralITIS products that the god-kings would have delivered.

These new scribes, their Homo Sap Thought-processing imperative directing them, began to produce WritingITIS products with individual IT styles for configuring and simplifying the hieratic WritingITIS elements. They began to capture Thought structures that they sequenced in the free space of their individual Homo Sap memories.

The Thought structures contained IS components disseminating messages describing inter-relationships between individual Homo Saps and between individual Homo Saps and their gods. They gave their gods different names, but all with attributes and characteristics of one or more of the nine cosmic gods, Geb and Nut, Ra and Thoth, and Osiris, Isis, Seth, Nephthys, and Horus. The new WritingITIS products—storytellers labelled them "literature" and their IS applications "fiction"— disseminated messages very different from those the god-kings and their priestly surrogates authorized.

More and more individual Homo Saps, as well as princely nomarchs, nobles, and those storytellers labelled as "middle class," began to practice their WritingITIS skills free of the conditioning of priestly and scribal institutional disciplines. They began to direct themselves with Thought structures they sequenced in the privacy of their individual Homo Sap memories. They captured and processed Thought structures from WritingITIS products that other individual Homo Saps produced, challenging those the institutions claimed as authorized by god-kings.

Homo Saps began to disrupt the orderly operation of the Ancient Egyptian civilization during the tenures of two god-kings. Storytellers gave them the names Pepy I and Pepy II. Both had relegated their leadership responsibilities to their institutional leaders. Their Homo Sap subjects waited for the god-kings to fulfill their leadership obligations and provide them with adequate survival food resources. The god-kings, unable to divert their Thought processing from the god-king Thoughtbase to their Homo Sap leadership Thoughtbase, and relying on their institutional leaders to identify the production problems, failed to produce the production procedures that would offset the disastrous environmental conditions along the Nile River. Survival food production continued to deteriorate.

Some individual Homo Saps, their HAP survival imperative compelling them, began executing hominid animal killing procedures to eliminate the competition for the scarce survival food resources. They first attributed the causes of the disastrous Earthly environmental conditions to the god-kings and their surrogate institutions. They focused their hominid animal killing procedures on eliminating the god-kings and priests.

They began disassembling the Ancient Egyptian operating system, eliminating the Thought-processing controls of the god-kings and their surrogates, and freeing their HAP survival procedures from their Thought-processing restrictions. As their hominid animal ancestors had done in the past, individual Homo Saps focused their HAPOS on ensuring the survival of their individual HAP structures. They indiscriminately killed other Homo Saps who competed with them for survival food resources. The killing events introduced a new chaos.

Homo Saps maintained the chaotic OS in the safe havens for ten Homo Sap generations. Individual Homo Saps, mostly those with WritingITIS skills with which they had expanded their Thought-processing capabilities, sequenced Thought into structures with IS components disseminating messages proclaiming themselves legitimate god-kings, claiming blood-line connections with previous god-kings, They executed primarily hominid animal killing procedures to legitimize their claims, forcing individual Homo Saps to react as hominid animals and challenge the claims. Homo Saps labelled the outcome of these challenges as "civil war."

Storytellers highlighted the impacts of free individual ITIS skills on Homo Sap civilizations by distinguishing old OS from the new. They labelled the old OS as that of the civilization of the Old Kingdom when god-kings and their institutions exercised absolute Thought-processing controls. They labelled the modified OS as that of the civilization of the Middle Kingdom, the outcome of the exercise of individual ITIS freedoms.

They labelled the period between the destabilizing of the Old Kingdom OS and the establishment of the Middle Kingdom OS as the "first intermediate period." They described the period as one during which individual Homo Saps, mostly members of the aristocratic institutions who had acquired ITIS skills, challenged and wrested control from the god-kings and their institutions in attempts to establish themselves as god-kings. They applied their ITIS skills to convince their fellow Ancient Egyptians that they had the Homo Sap leadership skills to resolve the food production and destabilizing problems that the poor

Thought-processing capabilities of the old god-kings and institutions had caused.

At the end of ten Homo Sap generations of civil war, an individual Homo Sap sequenced and disseminated Thought structures whose procedures and IS components produced outcomes that demonstrated outstanding Thought-processing skills. He convinced his fellow Ancient Egyptians of his leadership capabilities.

They agreed to grant him Homo Sap leadership privileges and stopped processing only HAP survival Thought structures. They agreed to re-install a modified Ancient Egyptian OS in all 40 safe havens along the Nile River.

Storytellers gave the name "Amenensis" to this individual Homo Sap. The outcomes of his Thought processing demonstrated his skills as an ITISan, Homo Sap leader, and god-king. As had the consummate ITISan, Narmer, 50 Homo Sap generations before, the Homo Sap Amenensis sequenced Thought structures whose outcomes and IS components identified kingly authority, institutions, and a stable OS as necessary components of an orderly, stable Homo Sap civilization.

He declared himself god-king, but different from previous god-kings. The IS components of the Thought structures he sequenced to proclaim his god-king status omitted any messages asserting his absolute authority. Instead, the messages acknowledged the authority of the nomarchs and nobles, all members of the aristocratic institution, and their hereditary claims as independent Thought processors in their nomes, but under his kingly leadership. He re-established the institutions, again under his kingly leadership. He devised new procedures for survival food production, procedures that allowed the survival food production cadres to optimize their productivity in all Earthly environmental conditions. He sequenced his Homo Sap leadership Thought in structures whose IS components disseminated messages that assured all Homo Saps equal treatment in the applications of the legal utilities of his new operating system.

Amenensis adopted the operating system that Narmer had designed as the kernel of his new operating system. His Homo Sap leadership analysis identified the outcomes of several utilities as causes of ten generations of civil war. The original applications of the utilities failed to anticipate the independent Thought processing progress in Homo Saps who had acquired WritingITIS skills. He chose to make minor changes to some utilities and major changes to two of the most contentious.

He changed the utilities whose outcomes automatically imposed his absolute god-king powers to control all Homo Saps and events in the Ancient Egyptian civilization. The changes gave members of the aristocratic institution the freedom to sequence and process their own Thought structures challenging those he produced. When they differed, they had to agree on consensus outcomes, compromises that acknowledged his god-king authority and their right to satisfy the management requirements in their jurisdictions.

He devised a Thought-processing interface through which the members of the aristocratic institution accessed the kernel operating utilities that guaranteed order in the safe havens and the safety and equal treatment of Homo Saps wherever they chose to operate along the Nile River. He designed the interface as a programming interface that identified applications that were incompatible with the guarantee. Only god-kings maintained and controlled the kernel operating system to ensure that all Homo Saps who chose to operate along the Nile River followed the same basic operating procedures and acknowledged the new supervising hierarchies of god-kings, princely nomarchs, and nobles.

Amenensis' Thought processing exhibited a high ITISanian skill for balancing his Homo Sap leadership responsibilities with his god-king responsibilities. He first changed the kernel utility whose outcome enforced god-king authority over all implementations, applications, and products of the four ITIS tools. His Homo Sap leadership Thought processing directed him to modify the outcome to one that freed individual Homo Saps to process any Thought structures they sequenced in the privacy of their individual memories and capture them in ITIS products.

The freedom allowed them to expand their Thought-processing capabilities, a benefit satisfying his leadership obligations to enhance their individual HAP survival capabilities. His ITISanian Thought processing identified a practical outcome to this freedom. Nomarchs and other members of the aristocratic institution operated in and managed safe havens along the Nile River with different Earthly environments. The uncertainties of the EOS could arbitrarily change environmental conditions, making standard god-king authorized operating procedures inefficient or ineffective. By releasing them from total god-king Thought-processing controls, he gave them the individual authority to devise new procedures and implement them without first seeking god-king approval.

He did impose one restriction on this freedom. He prohibited them from making any changes in ITIS tool specifications. His ITISanian Thought processing indicated that the use of the same versions of the ITIS tools, particularly OralITIS and WritingITIS, promoted clear communications between his Homo Sap subjects wherever they chose to operate. Clear communications—an automatic outcome of using the same IT and communications protocols—was necessary for the establishment and maintenance of the cohesion and unity of his Homo Sap civilization.

His ITISanian analyses also directed him to an outcome resolving his god-king requirements for stability and control of Thought processing. The specifications of the ITIS tools contained subtle but powerful Thought-processing controls as well as maintaining unity. Narmer originally designed the ITIS tools to capture and disseminate IS components with messages that continuously broadcast the power of the gods and god-kings over their Homo Saps' HAP survival. He had declared the hieroglyphic symbols as sacred, each symbol the access key to a dedicated Thoughtbase that gods transmitted to the god-kings and their priests. Only god-kings and priests had access to the godly encryption codes that allowed them to capture and decipher the godly messages and directives. They implemented the communications protocols that allowed them to capture and retransmit, in the Ancient Egyptian version of OralITIS, the godly directives and messages embedded in sequences of hieroglyphic ImageITIS products. The communications protocols depended on strict adherence to ITIS specifications. Changes to the specifications posed the greatest danger to god-king authority and Thought-processing controls, a danger that influenced the priestly and scribal institutions' development of the hieratic version of WritingITIS.

The scribes were careful to maintain the connection of hieratic with hieroglyphic WritingITIS. They designed each hieratic ImageITIS product as a shortened, easily-constructed substitute for a specific hieroglyphic ImageITIS product or sequence of products. They developed separate subsidiary Thoughtbases in their individual Homo

Sap memories to contain only the godly Thought structures embedded in each hieroglyphic ImageITIS product. They used the simpler, user-friendly hieratic ImageITIS products as the keys that automatically retrieved the associated hieroglyphic ImageITIS products, which in turn accessed the associated godly Thoughtbase.

This ITIS process allowed them to both improve their productivity and maintain the attributes of hieratic WritingITIS. The process was one they had learned to implement as a HAP survival procedure. Whenever their HAP visual IT identified an ImageITIS product that signalled a danger to their HAP survival, the product triggered their HAP survival imperative to retrieve and execute survival procedures that neutralized or avoided the danger. Individual Homo Saps were to implement the ITIS process to simplify and make more user-friendly the ITIS tools throughout their Earthly development as Thought processors.

Amenensis' modifications to the absolute authority of god-kings and Thought-processing freedoms initiated several generations of peace and a return to prosperity in Ancient Egypt. The successes of the freedom to process and disseminate new Thought structures in ITIS products prompted one of his successors—storytellers gave him the name Senuseret I—to sequence and disseminate Thought structures whose IS components disseminated messages promoting and exhorting all Homo Saps to acquire ITIS skills. The messages encouraged individual Homo Saps to free themselves from the Thought-processing restrictions that god-kings had imposed on them in the previous 50 generations.

Senuseret demonstrated high Homo Sap leadership and ITISanian Thought-processing capabilities. Ignoring the dangers to his god-king authority, he actively encouraged individual Homo Saps to capture and process Thought structures that others besides god-kings disseminated in OralITIS and WritingITIS products. He promoted the ITIS activities—storytellers labelled them "reading"—and the constant practice of WritingITIS skills as primary sources of Thought structures whose IS components would help them improve or enhance individual HAP survival.

Storytellers labelled this method for the capturing and processing of Thought structures in individual Homo Sap memories as "learning." The outcome was a period that they describe as the classical era in the development of Ancient Egyptian ITIS products.

Homo Saps began producing ITIS products containing Thought

structures with IS components disseminating messages that would have earned them the wrath of their god-kings and institutions. They produced ITIS products that identified the inter-relationships between individual Homo Saps and between individuals and their gods. The IS components challenged institutional claims that only god-kings and the religious institutions had legitimate relationships with gods.

They began to produce WritingITIS products with Thought structures disseminating instructions on how individual Homo Saps should operate in the safe havens. More dangerously, they produced WritingITIS products with Thought structures that defined the traditional social contract all Homo Saps had with the Homo Saps they chose as their leaders and what they should expect from them in fulfillment of the contract.

This Thought-processing freedom allowed them to modify the IS components of ImageITIS products. Individual Homo Saps designed statues of their god-kings and Homo Sap leaders to contain Thought structures with IS components disseminating messages identifying them as hominid animals with deteriorating HAP structures, the outcomes of the recycling EOS and COS utilities inexorably destroying the physical structures of all animals as they progress through their Earthly life cycles. For the previous 25 generations when the god-kings exercised absolute control of the production of ImageITIS products, all statues of the god-kings depicted them as massive idealized figures, perfectively formed, and unchanging.

This ITIS productivity and freeing of individuals from Thought-processing restrictions initiated many modifications to the civilization's OS. The outcomes demonstrated the power that free individual ITIS skills can have on the development of Homo Sap civilizations, a power that institutions would try but eventual fail to curtail in the development of all future Homo Sap civilizations.

The growing widespread use of ITIS skills spawned two more periods of disruptions in the Ancient Egyptian OS. Storytellers labelled them the "Second and Third Intermediate Periods." In between these periods, the god-kings re-stabilized the OS. Storytellers labelled the stabilized system as that of the "New Kingdom," during which the god-kings and their institutions were again paramount. After the New Kingdom, Homo Saps belonging to other groups—Mesopotamians and Ancient Greeks—invaded and took physical control of the whole or parts of Ancient Egypt.

The invaders failed to have any lasting impact on the OS kernel that Narmer had developed to help form the Ancient Egyptian civilization. They adopted the core operating procedures and adapted their lifestyles to that of the Ancient Egyptians. They left unchanged Narmer's core specifications of the ITIS tools and hence the Thought-processing controls he embedded in their IT and IS components.

The Homo Sap leaders that the Ancient Egyptian Homo Saps accepted and tolerated continued to claim godly status. They proclaimed their individual exclusive right to communicate with the gods and receive godly Thought structures for directing their supervisory control of operations in the safe havens along the Nile River.

A measure of the Thought processing freedom that Homo Saps, whether noble born or commoner, had wrested from their god-kings and institutions was the freedom to worship individual gods and naming offspring after gods. They made naming individual Homo Saps an ITISanian process. Parents, following the examples of their god-kings, gave their offspring a name corresponding to an epithet of a god, hoping that the offspring acquired the attribute the epithet specified.

The god-kings and priestly institutions, fearful of losing their exclusive control of communications with the gods, began to disseminate Thought structures with IS components claiming that only they had access to the true names of the gods. The messages in the IS components proclaimed the godly names as hidden in religious WritingITIS products. The messages claimed the gods themselves prohibited the dissemination of their true names. They prohibited individual Homo Saps from acquiring, identifying, and processing their true names in their individual Homo Sap memories, fearful of individual Homo Saps acquiring and implementing the godly powers embedded in the names,

Storytellers compiled their WritingITIS products with Thought structures describing the hominid animal killing actions between Ancient Egyptian Homo Saps and invading Homo Sap groups and the ITIS product developments, particularly ImageITIS, of the god-kings. They associated the greatness of god-kings with the battles and wars they won or lost and erected ImageITIS monuments to proclaim their greatness.

The increasing freedom to sequence and process Thought structures independently in individual Homo Sap memories prompted many individuals to free their Thought-processing imperative from godly and institutional restrictions. They produced WritingITIS products packaging Thought structures with IS components disseminating a wide range of new messages. Many messages focused on describing new individual relationships between Homo Saps and between individuals and gods. They named many new gods, each exhibiting a godly attribute that would resolve a problem or protect them from a danger.

Some messages proclaimed the return of the cosmic gods Ra and Thoth to the direct control of Earthly events. God-kings gave themselves names that proclaimed their direct connection them with Ra or Thoth and their Earthly powers. Priestly Homo Saps formed themselves into a cult that revived Thought processing of the nine original Ancient Egyptian cosmic gods.

The priestly Homo Saps identified the nine gods in WritingITIS products that packaged Thought structures describing their powers and actions and the godly impacts on the development of Homo Saps as Earthly entities. The priestly Homo Saps hid the WritingITIS products, claiming that Homo Saps had yet to the develop the Thought processing skills capable of processing the godly Thought structures and implementing their procedures without wreaking heavy damage on HAP structures or their Earthly environments. They kept well hidden the WritingITIS products that, they claimed, Thoth had directed them to compile and hide, for they described Thoth's consummate ITISanian and Thought processing skills—skills that would give the power of gods to Homo Saps and trigger the Atumic anger of Ra, the supreme god of the Solar System. They labelled the outcomes of those skills "wisdom and magic."

Ra had forced himself to ignore events in the Earthly environment since his confrontation and eventual agreement with Thoth hundreds of generations before Narmer initialized the Ancient Egyptian Homo Sap civilization. His inability to rid the Earthly environment of his hominid clones left him Atumically frustrated. His unAtumic Thought processing, primarily the outcome of Thoth's stimulation, convinced him that his only alternative—the destruction of Earth—left him vulnerable to the supreme cosmic god Atum's wrath. He still conditioned his Thought processing through the filter of his Atumic interface.

He began to refocus on Earthly events after he had been receiving Thought structures from his clones. The IS components transmitted a confusing set of messages. They indicated that his hominid clones were acknowledging a plethora of godly entities, overriding his godly admonition warning them of the consequences of their acknowledging gods other than him as supreme god.

The messages triggered his Atumic anger but his unAtumic Thought processing warned of the Atumic dangers and futility of using his Atumic powers to make the hominids extinct. Some had always survived and thrived, though his past violent Atumic actions had devastated and made extinct other species of his Earthly clones.

He processed his unAtumic Thought processing outcomes through the filter of his Atumic interface. His Atumic caste refused to accept that his hominid clones had acted independently to defy his directives. They had to have had godly help. He dismissed any actions of his five younger Earthbound siblings since they had chosen to confine their godly powers or Thought processing only to outcomes possible within the physical limitations of their HAPOS. His godly parents, Geb and Nut, left him alone, as did the supreme cosmic god Atum, with the Atumic threat of interfering only if his actions in the Solar System disturbed the equilibrium of the surrounding stable Cosmos. The only other cosmic god who had any possible interest in influencing Earthly events was his younger brother, Thoth.

Always ready to develop a legitimate Atumic excuse to rid himself of Thoth's aggravating presence in the Solar System, he accessed the Thoughtbase detailing his previous encounters with Thoth. He reprocessed the Thought structures he had sequenced to describe Thoth's actions in their confrontations. Thoth had succeeded in helping their mother Nut birth their five siblings and giving them Earthly sanctuary by tricking him and influencing his unAtumic Thought processing. The outcome had had little impact on the Solar System. He chose to use Thought processing rather than threats to their HAP structures to re-establish his godly suzerainty.

He transmitted Thought structures to all his Earthly god-king surrogates, all of whom, whichever Homo Saps group they led, asserted his suzerainty as Sun god. He included in the Thought structures procedures directing the Ancient Egyptian god-king to whom storytellers gave the name "Amenhotep IV" to dismantle the existing priestly institutions and initialize a new priestly institution with Thought structures whose IS components disseminated messages proclaiming him, the Sun god, as the one and only god. He chose the Sun disk that storytellers labelled "Aten" as the symbol of the new institution, and instructed the god-king Amenhotep IV to rename himself Akhenaten.

The newly-named god-king Akhenaten executed the procedures upon receiving and processing the cosmic god Ra's Thought structures. Akhenaten forced all his Ancient Egyptian Homo Saps to process the messages proclaiming the cosmic god Ra, the Sun god, as the one and only supreme god. He designed ImageITIS products—a new safe haven, temples and statues—dedicated to the Sun god and proclaiming the new godly order. Akhenaten failed to change the specifications of the other ITIS tools to capture and disseminate IS components and messages proclaiming the suzerainty and absolute authority of a single god.

All other ITIS products, as they had done for generations, continued to proclaim the cooperative of other gods. Akhenaten was unable to suppress their centuries-old messages. When he died, the older religious and aristocratic institutions whose powers and authority he had forcibly denied, ignored his directives proclaiming Ra as supreme god and reinstated the older institutional operating systems. They abandoned the safe haven dedicated to the Sun god and attempted to destroy all ImageITIS products proclaiming Akhenaten the Earthly surrogate of the Sun god. They defaced his statues and began to expunge all references to him in WritingITIS products. The priestly institutions resumed the rituals asserting the Earthly powers of other gods.

This action alone initialized a new message that individual Homo Saps of all future generations were to keep expanding and testing. The message proclaimed their independence as unique individual Homo Saps, free to choose for themselves the gods to whom they paid homage without fear of the sanctions of other gods.

Ra and Thoth had to process and reprocess the Thought structures describing these Earthly events. Both had failed to anticipate the outcomes. Both had identified the subtle change in Homo Saps' Thought processing about their gods.

The focus of Ra's analysis was how to regain control of their Earthly actions and outcomes. His unAtumic analysis of his hominid clones' Earthly development and his past failures in eradicating them with Atumic actions convinced him to use their Thought-processing capabilities to achieve his ultimate outcome. He chose to keep disseminating Akhenaten's Thought structures proclaiming the absolute authority of one supreme god.

Storytellers claim he designed his strategy as a challenge to the Earthly powers of other gods. He chose as his Earthly champions a Homo Sap group which already claimed to pay homage to only one god. They labelled the group "Hebrew" and gave the name "Yahweh" to their god. He chose as his primary Earthly champion a Homo Sap to whom they gave the name "Moses." He focused his Thought structure transmissions on Moses until Moses' only option to relieve the godly pressure on his Thought processing was to confront and openly challenge the reigning god-king who kept the Hebrew Homo Saps in slavery. Storytellers gave the name "Ramses" to the god-king.

Moses threatened Ramses with his god's anger if he failed to free the Hebrew Homo Saps and let them leave Ancient Egypt. Moses demonstrated the powers of his god, but Ramses was able to get his priests to duplicate most of the tricks. As a final threat, Moses threatened to get his god to initiate a plague that would kill all first-born sons in Ancient Egypt. When Moses' god succeeded in the threat, Ramses at first agreed to let the Hebrew group leave Ancient Egypt, acknowledging the overwhelming power of Moses' god. After the Hebrew group left, Ramses chose to chase them and kill them before they crossed a sea to safety.

The storytellers who compiled the Thought structures describing the confrontation between Moses and Ramses belonged to the priestly Hebrew institution. They labelled the sea the "Red Sea." They described the outcome as a further demonstration of their god's Earthly powers. Their god used his powers over the Earthly OS to divide the Red Sea into two parts with a fordable channel in between to allow the Hebrew Homo Saps to escape. As Ramses' army followed them into the channel, he released his control of the Earthly OS, allowing the sea to flow back into the channel to drown Ramses' soldiers and their chariots.

Ra was successful in maintaining and disseminating the Thought structures describing the omnipotence of a single supreme god. The Hebrew Homo Saps survived and developed an OS with utilities whose outcomes focused on the messages in the Akhenaten Thought structures. Priestly Homo Saps of future groups—particularly those who labelled themselves "Christians"—adopted and disseminated the single-god OS, accomplishing the outcome Ra had originally set for himself.

The focus of Thoth's analysis was how to keep Homo Saps from misusing their Thought-processing freedom. Thoth accepted Homo Saps new status as independent Thought processors but was wary of the possible outcomes. He transmitted Thought structures to Homo Saps who formed themselves into a priestly institution dedicated to promoting his Thought-processing achievements. Storytellers labelled the outcomes of the Thought structures he had sequenced from MEMORY as "wisdom and magic."

Thought structures specified procedures for individual Homo Saps to access MEMORY for Thought structures whose IS components described wisdom as a godly attribute that individual Homo Saps could acquire with Thought processing. The Thought structures also defined procedures for producing outcomes that to them appeared magical in their real Earthly environment. Thoth instructed his priestly Homo Saps to hide the WritingITIS products until all Homo Saps had developed individual Thought-processing facilities capable of applying his wisdom and magic Thought structures without destroying themselves or their Earthly environment.

Thoth initiated one other ITIS development to assist all Ancient Egyptian Homo Saps in acquiring the level of independent Thought processing that would allow them to execute his wisdom and magic procedures. He prompted scribes to develop a new, simpler version of Ancient Egyptian WritingITIS. Storytellers labelled this new version of WritingITIS "demotic." The scribes failed to accomplish Thoth's outcome. They produced the demotic version as a modification of hieratic, which was fundamentally a modification of hieroglyphics. They were unable to neutralize the Thought-processing controls or god-king messages embedded in the design of the original hieratic/hieroglyphic ImageITIS elements of Ancient Egyptian WritingITIS. They were unable to separate demotic from 100 generations of hieratic and hieroglyphic sacerdotal tradition.

The Ancient Greek Homo Saps, 25 Homo Sap generations after that of the god-king Akhenaten, sequenced sets of Thought structures whose IS components disseminated messages unequivocally asserting the independent Thought-processing capabilities of individual Homo Saps and their freedom from the absolute controls of gods or institutions. They developed the user-friendly ITIS tool of AlphabetITIS—a Homo Sap development that had roots in the ITIS developments of Mesopotamia. AlphabetITIS gave individuals the freedom to capture their own Thought structures in WritingITIS without the godly or god-king messages embedded in the specifications of the ImageITIS elements of all previous WritingITIS tools. For all future generations, the gods through their Earthly surrogates, the priestly institutions, were to viciously prohibit individual Thought processing and the development of ITIS products that disseminated messages different from those of the gods.

8:

ITIS in Mesopotamia

Dear Jim:

Storytellers labelled "Sumerian" the Homo Sap group that established safe havens along the lower stretches of two rivers during the same Homo Sap generations as those of the first Ancient Egyptian Homo Saps. They labelled the two rivers, the "Tigris" and the "Euphrates," and the Earthly region through which the rivers flowed as "Mesopotamia."

The storytellers also compiled sets of Thought structures with IS components describing a provenance for the first of the Sumerian Homo Saps who began to establish the safe havens. The messages in the IS components indicated that the ancestors of the Sumerian Homo Saps had to complete critical ITIS processes before they coalesced into a Homo Sap group with characteristics distinctive enough to allow the storytellers to give them the unique label of Sumerian. Executing these processes was critical to their development of an OS capable of sustaining their HAP survival in the new operating environment of the river region.

Storytellers themselves had to execute an ITIS utility whose outcomes they labelled as "speculation" in order to specify the provenance. The input parameters of this ITIS utility were Thought sequences they captured from objects they label as "artefacts": the physical remains of products that Homo Saps designed, manufactured, and used in safe havens before Earthly events completed the disintegration of the safe havens and the physical products into their natural Earthly components during the EOS recycling procedures. They processed the artefact Thought sequences in the free space of individual Homo

Sap memories, attempting to connect them into Thought structures with IS components and outcomes that speculated on how the Homo Saps who originally manufactured the objects operated or survived in the Earthly region. Their speculations and Thought structuring from Sumerian artefacts produced the outcome that the Homo Saps who coalesced to form the Sumerian group originated in a civilization they labelled "Saharan."

The Saharan Civilization

Storytellers classified the version of the OS that controlled the Saharan civilization as "gylanic." They defined a gylanic civilization as one in which the OS controls ensured that all interactions, whether in physical activities or Thought processing, between individual Homo Saps should be without conflict and that all Homo Sap groups have equal status in the operation of the civilization.

A key storytelling Thought structure contains an IS component that identifies a Thought-control feature that OS of other versions of Homo Sap civilizations allow Homo Saps to implement. Homo Saps labelled the outcome of the utility as "dominance."

Gylanic operating procedures contain embedded IS components disseminating messages prohibiting the implementation of the Thought-control feature, which allows one group of Homo Saps to dominate another. The procedures directed individual Homo Saps or Homo Sap groups to inhibit activities whose outcomes gave them dominance over other individual Homo Sap groups which operated in the gylanic safe havens.

EOS events forced the Saharan Homo Saps to change their OS. The EOS events reduced the productivity of the survival resources sustaining Homo Saps operating in the Saharan Earthly region. Their HAP survival imperative compelling them, the Saharan Homo Saps had to migrate to Earthly regions with survival resources capable of sustaining their HAPOS. Some storytellers—those who operate exclusively in the classic Thoughtbase—classify the outcome of this migration as the beginning of existing Homo Sap civilizations and of their ITIS achievements, ignoring the impact of the Ancient Egyptian civilization.

Starting with Thought sequences they retrieved from Saharan artefacts, storytellers sequenced Thought into structures whose IS components described the Sumerian Homo Saps as offspring of one Saharan Homo Sap group. They completed the Thought structures with Thought sequences they retrieved from the OralITIS products that different Homo Sap groups developed after they had established their safe havens in Earthly regions outside the Saharan Earthly region: One as distant from Sumer as the region storytellers labelled as "Basque."

The storytellers described the OralITIS products of these different groups as having similar IT and IS components. Using these similarities to justify their outcomes, they speculated that the different Homo Sap groups were the offspring of Homo Saps who migrated from the Saharan gylanic civilization.

Implementing the ITIS utility of speculation, storytellers sequenced Thought structures that described the Saharan OS as one in which male Homo Saps and female Homo Saps operated in complete harmony. The male Homo Saps executed survival procedures directing them to travel from their safe havens to kill animal species whose physical structures contained components with Earthly elements that Homo Saps needed to sustain their HAPOS. The female Homo Saps executed survival procedures directing them to remain in the safe havens to ensure the physical survival of their Homo Sap offspring and to gather sustaining Earthly elements from plant species that the EOS allowed to survive and grow in the Earthly region close to the Saharan safe havens. The male Homo Saps returned to their safe havens with animal survival resources enough to sustain the HAP structures of all Homo Saps operating in the safe havens. When they needed more animal survival resources, the male Homo Saps again left the safe havens to fulfill their HAP survival responsibilities.

The Homo Saps who formed the Sumerian Homo Sap groups developed their safe havens on the banks of the lower reaches of the Euphrates and Tigris rivers. The prevailing EOS utilities of the river region had established a survival environment that was different from that of the Saharan environment. The differences made the survival applications of the gylanic OS inefficient in supplying them with survival resources.

The differences impacted primarily the HAP survival responsibilities of the male Homo Saps. The river environment sustained much fewer animals than the Saharan environment. The Homo Sap males had to develop new procedures and applications to define their survival roles in the river environment. The only immediate role was to develop the procedures of "husbandry" and "farming."

The husbandry and farming procedures clashed with those of the female Homo Sap survival responsibilities. The migration to the Mesopotamian region and the establishment of new safe havens had left virtually unchanged the gylanic responsibilities of female Homo Saps. They still had to maintain the operation of the safe havens, gather plant survival resources, and nurture their offspring.

The successful production of abundant survival resources from the farming and husbandry procedures required the male Homo Saps to remain in the safe havens. The need to change their Thought processing from hunting procedures to husbandry and farming procedures forced them to begin Thought processing in the free space of their Homo Sap memories. This was a fundamental ITIS event that impacted the Thought processing of only the male Homo Saps. The female Homo Saps still operated in their gylanic Thoughtbase.

Male Homo Saps initiated the ITIS process that helped them implement a new OS that forced fundamental changes in the interrelationships of male and female Homo Saps. To consolidate the new OS they had to redesign the IS features of the Saharan version of OralITIS that disseminated gylanic messages. They re-engineered the IT features to develop the new version of the ITIS tool of OralITIS. Storytellers labelled the new version as "Sumerian."

This ITISanian Thought processing initialized the ITISanian Thoughtbase containing Thought structures whose IS components described the procedures necessary for making fundamental changes in the OS of civilizations. The most powerful procedures were those that changed the design of the IT and IS features of the ITIS tools that the Homo Saps who operated in a civilization used to express their Thought-processing outcomes. Achieving the change using HAP killing procedures invariably failed. The Homo Sap leaders who produced the most successful changes were those who focused on the control of ITIS tools and their applications rather than HAP killing procedures to eliminate those Homo Saps who opposed changes.

The Saharan Homo Saps took more than fifty generations to complete the design and implementation of the Sumerian version of OralITIS. They replaced and eliminated all applications of the Saharan version of OralITIS, erasing IS components from OralITIS products whose messages conditioned Homo Sap Thought processing to operate in gylanic civilizations.

The Sumerian OralITIS products contained IS components conditioning Homo Sap Thought processing to comply with the new procedures and outcomes of the new Sumerian OS.

At their completion of the ITIS event and establishment of the outcomes, Homo Saps initialized ITISanian Thought structures that they periodically retrieved, reprocessed, and augmented in every Homo Sap group that they established after the Sumerian Homo Sap groups established themselves. They embedded IS components with messages describing the ITIS tools as the most effective in redesigning and successfully establishing new OS for safe havens.

The Sumerian OS was radically different from the Saharan OS. The Sumerian Earthly region had different survival and food resources than those in the Saharan Earthly region. The Thought structures defining the new OS contained IS components whose messages described the power of individual Homo Sap Thought processing in individual Homo Sap memories in designing and implementing ITIS tools and changes to ITIS tools to initiate, reinforce, and ultimately establish new OS whose procedures allowed Homo Saps to operate and survive in new Earthly environments. These ITISanian Thought structures and messages were to direct the Thought processing of the most successful leaders of the Mesopotamian Homo Saps.

URAKAGINA, SARGON, AND HAMMURABI

Three Mesopotamian Homo Sap leaders sequenced and disseminated the Thought structures specifying the procedures that formed the kernel of the OS of the Mesopotamian civilization.

Storytellers gave the name "Urakagina" to the first, a member of the Sumerian Homo Sap group. Urakagina initialized the Thoughtbase of civilizing Thought structures. They gave the name "Sargon" to the second, a member of the Homo Sap group that Homo Saps labelled "Akkadian." Sargon added to and expanded Urakagina's Thought structures.

Storytellers gave the name "Hammurabi" to the third of the three Homo Sap leaders. He was a member of the Homo Sap group that they labelled "Amorite." Hammurabi modified, added to, and expanded Urakagina's and Sargon's Thought structures to fix the specifications for the OS kernel that finally established a Homo Sap civilization of the groups which chose to operate along the banks of the Tigris and Euphrates rivers.

Hammurabi consolidated the Mesopotamian civilization when he developed and disseminated fixed specifications for the versions of the four ITIS tools of OralITIS, ImageITIS, CalendarITIS, and WritingITIS that he chose to use to control his disparate groups of Homo Saps. His ITISanian analysis of 50 Homo Sap generations of events in the safe havens along the two rivers indicated that fixing ITIS tool specifications was essential for the stability and cohesion of any Homo Sap civilization. He forced all Homo Saps who chose to operate in the safe havens to implement and comply with the specifications of his ITIS tools. Hammurabi designed his OS to incorporate the Thought structures defining the most effective operating procedures and the IT and IS components of the ITIS tools of the original Sumerian and Akkadian Homo Sap groups.

Hammurabi's ITISanian analysis identified the stability and cohesion problems of the generations before his. Different groups of Homo Saps from Earthly regions outside that of Mesopotamia entered and occupied the safe havens, displacing and replacing those Homo Saps whose institutions controlled survival resources. Each Homo Sap group processed Thought structures in Thoughtbases whose procedures and messages directed them to achieve outcomes different from those of the Thoughtbases controlling the safe havens they forcibly occupied. Their primary directive on first entering the safe havens was to comply with their HAP survival imperative. They executed the killing procedures that the imperative specified, disintegrating the Earthly physical structures of large numbers of the Homo Saps who supervised the operations of the safe havens. They established their control of the survival resources of the safe havens.

After securing their HAP survival, the Homo Sap leaders of each group in turn immediately complied with the directives of their Homo Sap Thought-processing imperative. They established their control of the ITIS tools of the safe havens. They implemented their versions of the ITIS tools of OralITIS and of WritingITIS to impose on all Homo Saps their procedures and protocols for Thought processing. They sequenced and disseminated new messages proclaiming their institutional dominance of the safe havens. They implemented the specific features of their ITIS tools to re-sequence Thought structures and modified outcomes contained in WritingITIS products that institutions gathered and maintained in ITIS products they stored in ITIS peripherals storytellers labelled "libraries."

SUMERIAN ITIS DEVELOPMENTS

Storytellers labelled as "Sumerian" the first of the Mesopotamian Homo Sap groups to develop the IT that allowed them to design their version of WritingITIS. The Sumerian Homo Saps designed implements with which to make marks on ITIS substrates that they made from Earthly materials available to them. They designed the marks as ImageITIS products in a unique configuration. When they captured the ImageITIS products with their HAP visual technology, the ImageITIS products acted as keys to automatically retrieve specific sets of Thought structures that they stored in individual Homo Sap memories and process them in their Homo Sap Thought-processing facility. The Thought structures described the purpose of the ImageITIS products and the procedures for implementing their purpose.

Some storytellers labelled the ImageITIS products as "pictograms." The first pictograms represented real objects. The Sumerian Homo Saps combined these ImageITIS products in specific sequences and scribed them on WritingITIS substrates. They processed the Thought structures in Homo Sap memories and joined them in specific sequences to produce new, more complex Thought structures. When they first developed these ITIS procedures, the new Thought structures described real products they manufactured for food and other survival resources that they grew for their own HAP survival or for exchanging survival resources with other Homo Sap groups.

The Sumerian Homo Saps originally designed their WritingITIS elements to capture Thought structures with precise outcomes and specific IS components for communication with specific groups of Homo Saps. Storytellers labelled the primary survival activities of these specific groups of Homo Saps as "trade and business." The functions that the Homo Sap groups which gathered in the safe havens assigned to these businesses focused on the provision of abundant survival resources for all Homo Saps who operated in the safe havens. Their HAP survival imperative directing them, the Sumerian Homo Saps designed their first version of the ITIS tool of WritingITIS with one outcome: the order, control, and production of abundant, sustainable survival resources.

The Sumerian Homo Saps stored the Thought structures in individual Homo Sap memories in Thoughtbases they dedicated to the procedures of trade and business transactions. The individuals responsible for business transactions focused their Thought processing on maintaining and refining the Thought structures. With their constant processing and reprocessing of the Thought structures, their Homo Sap Thought-processing imperative directed them to expand the Thoughtbases with Thought structures that specified the protocols and procedures for completing all business transactions with outcomes providing them with surplus survival resources. Storytellers labelled the surplus survival resources as "business profits."

The security of the safe havens allowed individual Homo Saps to increase the frequency of their HAP sexual activities and consequently increase their numbers, a HAP survival imperative. The sustenance of their increased numbers demanded more survival resources. This demand prompted their collective HAP survival imperative to assign the responsibility for the production, control, and management of their survival resources to the dominant institutions that maintained order and ensured the safety of individuals operating in the safe havens. Storytellers labelled the two dominant institutions as "aristocratic" and "priestly."

The Thought structures that contained IS components with messages justifying this assignment were ones the institutions themselves sequenced and disseminated. These Thought structures described extra-Earthly entities that storytellers labelled "gods," entities that had the omnipotence to control all Earthly events and the omniscience to direct Homo Saps in survival procedures and the production of survival resources. The institutions designed the IS components to disseminate messages identifying the aristocratic and priestly Homo Saps as the Earthly representatives of the godly entities and Earthly executors of survival directives that the godly entities transmitted exclusively to them in godly Thought structures.

Individual Homo Saps gave the members of the priestly institutions freedom from all physical survival activities. In return, the priestly Homo Saps agreed to sequence Thought into structures with IS components containing messages they transmitted to their gods and to capture Thought structures their gods transmitted to Homo Saps. They processed the godly Thought structures and disseminated any messages and survival procedures to Homo Saps operating in the safe havens.

Their freedom from physical survival activities allowed some priestly Homo Saps to acquire great skills in the applications of their version of the ITIS tool of WritingITIS, and Homo Saps assigned them the responsibility for maintaining WritingITIS products describing business transactions. Free to process Thought structures focused on business transactions, the demands of their Homo Sap Thought-processing imperative impelled them to expand business Thought structures to include procedures for storing, controlling, and distributing survival resources.

Most of the Homo Saps who joined the priestly institutions complied with the rigid Thought processing rules of the institutions. They confined their Thought processing to Thoughtbases they stored in their individual Homo Sap memories. The Thoughtbases contained all the Thought structures that specified the protocols and procedures for transmitting supplicant messages to their supervising gods and for receiving godly Thought structures with their directives and messages from the godly entities.

The priestly institutions developed and implemented applications of the Sumerian versions of the ITIS tools of OralITIS and ImageITIS to demonstrate to the other Homo Saps operating in the safe havens that they were executing their Thought-processing obligations. They sequenced Thought structures with procedures specifying Earthly physical activities that they executed while they were processing godly Thought. Storytellers labelled these activities "rituals."

The priestly Homo Saps coordinated the physical activities of the rituals with IT features of the OralITIS and ImageITIS products they produced. They conditioned their embedded HAP vocal IT to produce sequences of sounds that storytellers labelled "chants." They designed OralITIS products with Thought structures whose IS components disseminated messages that storytellers labelled "prayers." They combined the prayer Thought structures with sequences of sounds to produce OralITIS products that storytellers labelled "hymns." They described the sequences of words, sounds, and tones in the chants and hymns as those that the gods directed them to produce as carriers of their supplications and messages to the gods.

They designed ImageITIS products as visual representations of the Earthly presence of their gods. Storytellers labelled one set of ImageITIS products "idols," representing individual gods. They labelled another set of ImageITIS products "temples," representing the Earthly homes of the godly entities. The priestly Homo Saps designed OralITIS products with messages in which they explained to other Homo Saps in their safe havens the purposes of their physical prostration before the idols in the temples or their execution of other activities of the rituals.

Though priestly Homo Saps were members of an institution that dedicated their Thought processing to the fixed Thought structures of their institutional Thoughtbases, they were also individual Homo Saps with independent Thought-processing facilities and subject to the Homo Sap Thought-processing imperative. Their processing of institutional Thought structures automatically expanded each individual Homo Sap memory, an outcome all Homo Saps are unable to avoid after they begin Thought processing.

Each had the freedom to sequence Thought in the free space in their individual Homo Sap memories that continuous Thought processing expanded. Each had free access only to the Thoughtbases that they stored in individual Homo Sap memories. Each had the freedom to sequence free-Thought or re-sequence institutional Thought structures in the free space of individual Homo Sap memories in absolute privacy, security, and secrecy.

One Homo Sap was unable to access the Thought structures that another sequenced in individual Homo Sap memories. The Homo Sap Thought-processing imperative compelled them to process free-Thought into structures in the expanded space of individual Homo Sap memories, expanding each individual Homo Sap memory still more.

Priestly Homo Saps assumed that the gods transmitted the free-Thought to them when they prostrated themselves in the silence of temples before their godly idols. They claimed that the success of the ritual was confirmation of their special status in the safe havens as the chosen Earthly representatives of their gods.

Free-Thought-processing in individual Homo Sap memories always produced new Thought structures with outcomes and messages in IS components different from those of the original Thought structures. Assuming the new Thought structures were those that their godly entities prompted them to produce, the priestly Sumerian Homo Saps implemented the procedures that the Thought structures specified and disseminated their messages in their IS components.

These outcomes and messages caused disruptions in the operating systems and procedural discipline of the institutions. The disruptions prompted other Homo Saps to begin to lose confidence in the capability of the institutions to meet their HAP survival obligations.

To recover, the priestly institutions arbitrarily sequenced new Thought structures specifying prohibitions on free-Thought-processing. They connected the Thought structures to the Thoughtbase of institutional Thought controls. Their messages threatened retribution from both institutions and gods on those Homo Saps who failed to abide by the prohibitions.

Their HAP survival imperative directing them, the priestly Homo Saps suppressed any Thought structure that produced outcomes and messages different from those that their institutions established as standard. Unable to suppress their Thought-processing imperative, the priestly Homo Saps who functioned as scribes focused their Thought processing in the free space of their Homo Sap memory networks on modifying and improving the design of their version of the ITIS tool of WritingITIS. They expanded applications focused on the control of survival resources and business transactions that they were responsible for implementing.

The priestly Homo Saps began to design and develop thousands of new ImageITIS products to capture the IS component of new Thought structures. Storytellers labelled these ImageITIS products "ideograms." Their productivity marked an ITIS event whose outcome established the unlimited capability for sequencing and processing Thought in individual Homo Sap memories when each individual was free of fear.

In most animal species, the core ImageITIS products focused on retrieving survival Thought structures whose procedures directed individual animals in the physical activities they needed to execute in response to the immediate HAP survival needs that the ImageITIS products prompted. Storytellers labelled the connections between ImageITIS products, "survival Thoughtbases," and individual animal physical operating systems as "hard-wired," and the reactions to the ImageITIS products as "instinctive."

The priestly Sumerian Homo Saps demonstrated the great

productivity of individual Homo Sap Thought processing forty Homo Sap generations after the production of the first Sumerian WritingITIS products, when they freed themselves of the Thought-processing constraints of fixed institutional Thought-control networks. Their Thought-processing imperative impelling them, they sequenced more and more new Thought structures, often melding them in the free space of individual Homo Sap memories to produce more, and more complex, Thought structures. They needed new WritingITIS elements to identify the new Thought structures and designed new ImageITIS products to represent the new WritingITIS elements on ITIS substrates. During these ITIS developments, the priestly Sumerian Homo Saps developed thousands of new WritingITIS elements and designed unique ImageITIS products to represent them.

The uniqueness requirement in the designs of ImageITIS products for inscribing on ITIS substrates prompted the priestly Sumerian Homo Saps to initiate a major development in the use of ITIS tools as aids to individual Homo Saps Thought processing.

Homo Saps outlined significant features of real environmental objects as components of the ImageITIS products. By including pictograms of real environmental objects in the ImageITIS products, Homo Saps indicated that the Thought structures that the ImageITIS products prompted them to retrieve contained sequences detailing survival resource data related to the real objects.

The new Thought structures the priestly Homo Saps sequenced initialized procedures and outcomes that had little connection with real objects or Earthly events. Storytellers labelled the new Thought structures as "abstractions" whose outcomes identified "realities" that Homo Saps defined only in the privacy of individual Homo Sap memories.

They labelled these realities as "love," "hate," "enmity," "friendship," "danger," "fear," and "death." To retrieve these Thought structures the Sumerian priestly Homo Saps designed ImageITIS products without pictograms of real objects. They used lines and combinations of lines in unique configurations to produce the new ImageITIS products. Very often the purpose of the new ImageITIS products was to prompt further Thought processing of the abstractions to identify a reality to which they related.

The priestly Homo Saps' freedom to produce thousands of unique

ImageITIS products undermined the priestly institution's Thought-processing controls. Each ImageITIS product represented a Thought structure that individual priestly Homo Saps sequenced in the free space of individual Homo Sap memories. The IS components of the Thought structures contained different messages and disrupted coordinated Thought processing in the priestly Sumerian Homo Sap memories.

The Sumerian institutions re-established their controls by imposing a limit on the numbers of ImageITIS products to a few hundred. They directed the priestly Homo Sap scribes to reproduce only those ImageITIS products that the institutions authorized in lists they inscribed on ITIS substrates. They disseminated the Thought structures that each ImageITIS product represented in the Sumerian version of the ITIS tool of OralITIS and forced the priestly Homo Sap scribes to store the design of the ImageITIS products and associated Thought structures in individual Homo Sap memories. Though the institutions regained control of Sumerian ITIS developments, they were unable to restrict more momentous ITIS developments that the free-Thought processing initiated.

Within 20 generations—400 Earth years—of individual Thought processing after the production of their first WritingITIS products, the Sumerian Homo Saps, both priestly and others who had acquired WritingITIS skills, initiated an ITIS event that storytellers described as the most critical in the development of ITIS tools and Thought processing in individual Homo Sap memories.

Homo Saps began developing the procedures for capturing facsimiles of their OralITIS products in their WritingITIS products. Before this event, Sumerian WritingITIS products transmitted self-explanatory details of business transactions, inventories of goods, and accounts. The new procedures allowed individual Homo Saps to inscribe their OralITIS products—their conversations, announcements, instructions—on ITIS substrates and transmit them to other Homo Saps who operated outside the range of HAP vocal IT.

They designed ImageITIS products to represent the distinct sounds that they trained their vocal technology to make in the application of their version of OralITIS. They sequenced the ImageITIS products into structures that they designated as representative of OralITIS elements that storytellers labelled as "words." They sequenced the OralITIS word elements into more complex structures that storytellers labelled as "sentences."

The new ImageITIS products allowed them to inscribe the words and sentences on ITIS substrates and capture in WritingITIS products complete Thought structures that they retrieved from Thoughtbases they stored in individual Homo Sap memories. This simulation of the sounds of OralITIS products as an application of WritingITIS allowed any individual Homo Sap to disseminate complete Thought structures, each sequenced in individual Homo Sap memories, and transmit them to others without the use of OralITIS procedures.

Priestly Homo Saps designed new procedures for capturing from WritingITIS products complete Thought structures without using OralITIS applications. The new procedures specified the sequencing of the ITIS elements of words and sentences into combinations capable of reproducing Thought structures that individual Homo Saps sequenced in individual Homo Sap memories. The new procedures allowed Homo Saps to reproduce Thought structures containing exact representations of the OralITIS elements that individual Homo Saps utilized in disseminating the Thought structures with their versions of the OralITIS tool.

Storytellers labelled the process of sequencing the OralITIS elements as "grammar" and the procedures the "rules of grammar." They stored the Thought structures describing the syntactic process and the sequencing rules in Thoughtbases in individual Homo Sap memories.

Storytellers labelled as "reading" the combined processes of capturing the ITIS components and messages of complete Thought structures that they inscribed in WritingITIS products, of retrieving and applying the rules of grammar to the sequences of WritingITIS elements, and of processing the Thought structures in individual Homo Sap memories. Storytellers labelled as "learning" the process of executing the procedures that the new Thought structures contained, of acting on the messages contained in their IS components, and of implementing the outcomes they specified.

The ITIS process of learning after completing the ITIS process of reading was a modification of the ITIS process that Homo Saps designed with the OralITIS and ImageITIS tools when they established themselves as a distinct hominid animal species. Homo Saps delivered to other Homo Saps new Thought structures in OralITIS and ImageITIS applications. They demonstrated the physical activities of operating and survival procedures to other Homo Saps in ImageITIS applications.

To implement the learning process, all Homo Sap participants gathered in groups at single locations where they were close enough to hear the Homo Saps delivering the new Thought structures. They had to operate within the ranges of their embedded vocal and aural ITs of their HAP structures to capture the new OralITIS and ImageITIS products. Duplicating the OralITIS products in WritingITIS products allowed individual Homo Saps to capture the Thought structures in the ITIS process of reading in isolation and process the Thought structures in their individual Homo Sap memories without distractions from other Homo Saps.

Priestly Homo Saps designed new ancillary ImageITIS products as WritingITIS elements to aid individual Homo Saps in achieving the outcomes of the ITIS process of reading. They labelled the ancillary ImageITIS products as "diacritics." The absence or presence of the diacritics next to ImageITIS products prompted individual Homo Sap readers to start two ITIS processes. They could search the Thoughtbases they stored in their individual memories for Thought structures with IS components describing the purpose of the diacritics. They could use them as prompts for reproducing with their vocal technology the sounds of their version of OralITIS in the sequences of word and sentence combinations they inscribed in WritingITIS products to capture their OralITIS products.

Fifty generations of Sumerian Homo Saps allowed their priestly institutions to control the development of their versions of the OralITIS and WritingITIS tools. In addition to Homo Saps' HAP fear of physical annihilation, the priestly Sumerian institutions developed applications of ITIS tools as the second most powerful tools for controlling Homo Sap Thought processing in their safe havens. The Thought structures that the priestly institutions sequenced to justify their controls demonstrated the freedom and capabilities of Homo Saps to sequence Thought into structures that justified any outcome they chose. Other Sumerian Homo Saps were to capture and implement the procedures in Thought structures.

As starting nodes for the Thought structures, the priestly institutions used the Thought structures identifying their successes in developing and implementing the first WritingITIS applications to control trade and business and ensure the availability of abundant survival resources. They retrieved the Thought structures describing their godly entities as the ultimate HAP survival controllers and guarantors. They retrieved the Thought structures designating the priestly institutions as the Earthly representatives of the godly entities. They joined and processed these Thought structures.

The priestly institutions chose a precise set with IS components containing clear messages from the infinite number of Thought sequences that individual priestly Homo Saps produced from this Thought processing. They proclaimed the godly entities as the owners of the Earthly facilities for the production of survival resources since the godly entities controlled the productivity of the land and other Earthly components.

With their assignment as the Earthly representatives of the godly entities, the priestly institutions automatically also assumed the role of Earthly owners and managers of land and production facilities. As godly-authorized Earthly owners of survival facilities, the priestly institutions had the right to benefit most from their successes. All other Homo Saps in the safe havens had to comply with institutional Thought directives since they had godly provenance. They had to accept only those freedoms that the priestly institutions assigned to each individual Homo Sap in the safe havens.

The Sumerian priestly institutions failed to attribute their successes and capabilities to the freedom to function exclusively as Thought processors. They filtered new Thought structures and sequenced through their institutional Thoughtbase interface and conditioning all outcomes to comply with the fixed godly outcomes which made the Sumerian priestly institutions unable to associate the development and implementation of ITIS applications with the expansion of the free-Thought processing facilities in individual Homo Sap memories and the social changes in the social systems of the safe havens. They failed to make the connection between ITIS developments and social changes or the power of Thought processing in individual Homo Sap memories.

Other Sumerian Homo Saps besides the priestly Sumerian Homo Saps began to acquire skills in implementing WritingITIS applications. Most of these Homo Saps were members of the aristocratic institutions.

As with their assignment of the Homo Sap members of the priestly institutions as the primary Thought processors in the safe havens, the Homo Saps who operated in the safe havens assigned the Homo Sap members of the aristocratic institutions the responsibility for the HAP security of the safe havens.

In accepting this assignment, the Homo Sap members of the aristocratic institutions dedicated their Thought processing to sequencing and executing procedures for maintaining the physical integrity of the environment of their safe havens. They protected their own safe havens from other Homo Sap groups or from individual Homo Saps who threatened to damage the HAP structures of individual Homo Saps in the safe havens.

The aristocratic institutions interacted with the priestly institutions and implemented procedures whose physical activities and outcomes maintained the social order that the priestly institutions imposed on the safe havens. The aristocratic institutions interacted with the priestly institutions primarily to obtain godly sanction to execute any procedures whose outcomes reinforced the security and social order of the safe havens.

The aristocratic institutions were as successful as the priestly institutions in meeting their social responsibilities in the safe havens. They devised order and security procedures that they executed as automatically as they did the procedures of the HAP survival imperative.

The aristocratic Homo Saps duplicated the procedures of the priestly Homo Saps for producing WritingITIS products. The Thought processing that individual aristocratic Homo Saps practiced was simple. They focused on the physical procedures for inscribing exact copies of the ImageITIS elements of WritingITIS products on WritingITIS substrates. Most physical skills that individual Homo Saps acquired focused on the production of physical products, and any further application of the skills focused on improving the physical utility of the products.

The use of ImageITIS products as elements of WritingITIS products had other purposes besides that of producing physical products. The primary functions of the ImageITIS elements were to prompt individual Homo Saps to retrieve specific Thought structures from individual Homo Sap memories and to process the Thought structures in individual Homo Sap memories. The demands of the individual Homo Sap Thought-processing imperative produced the inevitable outcomes: Individual aristocratic Homo Saps who acquired WritingITIS skills began to expand their individual Thought processing capacities.

The critical and powerful message in the new aristocratic Thought structures was in the freedom from the Thought conditioning and filtering features of the priestly institution Thoughtbase. The aristocratic Thought structures described operating procedures and outcomes that challenged those that the priestly institutions imposed on all Homo Saps operating in the Sumerian safe havens.

One aristocrat—a skilled ITISan—promulgated a set of Thought structures with procedures and outcomes that discounted the absolute authority of the priestly institutions in the safe havens. The Thought structures contained the first message that identified the rights of individual Homo Saps when confronting their social institutions.

URAKAGINA

Storytellers described Urakagina as king of the safe haven they labelled "Lagash." Urakagina claimed that a god that storytellers named "Ningursu" was the source of his Thought structures and the godly authority sanctioning his promulgation of the Thought structures in opposition to those of the priestly institutions.

The ITIS outcome of the Homo Sap Urakagina's free-Thought processing and production of WritingITIS products was a change in the controls on Thought processing in the safe havens. The aristocratic institution, independent of the priestly institution, began implementing Thought structures disseminating messages that claimed the aristocratic institutions' had rights to more social benefits from the increased social stability and physical security that the aristocratic Homo Saps had engineered in the safe havens.

The social outcome was the implementation of growing conflict between the priestly and aristocratic Homo Saps over which institution was the primary controller of events and Thought processing in the Sumerian safe havens. Both institutions began to focus their ITIS applications and developments on disseminating Thought structures whose outcomes and messages countered one institution's claim of superiority over the other.

Both institutions sequenced Thought structures describing the disrupting social changes that the Homo Sap Urakagina's free-Thought sequencing and processing initiated outside the controls of the institutional Thoughtbases. Focused on their conflict over institutional hegemony, both institutions fixed their controlling Thoughtbases and inhibited other Homo Saps from initiating ITIS developments whose outcomes threatened their godly and social authorities.

Both institutions promulgated Thought structures whose messages promised godly retribution on Homo Saps who used ITIS tools to threaten the physical survival of the institutions or to diminish the authority and hegemony of the institutions. Their fear of godly retribution and for their HAP survival prompting them, Sumerian Homo Saps stopped all further ITIS developments.

The fifty generations of Homo Saps up to that of the Homo Sap Urakagina had initialized and established all the features of the ITIS tool of WritingITIS except one: a user-friendliness that allowed individual Homo Saps to acquire ITIS skills. The requirement to capture thousands of fixed ImageITIS and OralITIS products and their specified Thought structures and store them in individual Homo Sap memories was un-user-friendly. In those fifty generations of Thought processing, the Sumerian Homo Saps demonstrated the power of the ITIS tool of WritingITIS in initiating Homo Sap Thought processing in individual Homo Sap memories to produce outcomes that benefited individual Homo Saps.

SARGON

Sargon initiated further developments of the ITIS tool of WritingITIS that the priestly Sumerian Homo Saps established. Sargon was the leader of the military institution of a distinct group of Homo Saps that storytellers labelled as "Akkadian" and made himself king of the Akkadians, the supervising Homo Sap of both the Akkadian military and aristocratic institutions. The Akkadian Homo Saps operated in safe havens that they established on the banks of the rivers Tigris and Euphrates, the same rivers that the Sumerian Homo Saps used as the source of their survival water resource. They had constructed their safe havens four hundred kilometers downstream of the Sumerian safe havens.

Sargon directed and led the Akkadian military institution in the occupation of the Sumerian safe havens and made the Akkadian Homo Saps the dominant Homo Sap group. Storytellers labelled an Earthly region in which two or more Homo Sap groups with distinct ITIS tools operate in a social system that one Homo Sap group dominated and controlled as an "empire." The Homo Sap Sargon established the first Homo Sap Empire that storytellers labelled the "Mesopotamian Empire."

Storytellers Thought structures described him as a powerful, successful leader of his military institution, highly skilled in the execution of hominid animal killing and domination procedures. His subjugation of the Sumerian Homo Saps was a hominid animal survival action. He coveted for his own Akkadian Homo Saps group the survival resources of the Sumerian Homo Saps group.

Storytellers also described Sargon as a Homo Sap leader who demonstrated Thought-processing capabilities that established him as an ITISan. Once he had made his Akkadian Homo Saps the dominant group, his ITISanian Thought processing directed him to analyze the Sumerian OS and the ITIS tools for the applications and features that helped the Sumerian Homo Saps make themselves more successful than his Akkadian Homo Saps in survival resource production. He focused first on ITIS features that helped ensure stability and his group's continued dominance in his Mesopotamian Empire.

The Akkadian Homo Sap group depended primarily on the dissemination of their OralITIS products to maintain control of the safe havens they dominated. The Sumerian Homo Saps had been developing their ITIS tools for fifty generations before Sargon. His ITISanian analyses of Sumerian ITIS features produced the outcome that WritingITIS was a much more powerful tool than either OralITIS or hominid animal killing procedures for stabilizing and controlling operations in safe havens. The key feature was the uniformity of Sumerian ITIS products. They forced Sumerian Homo Saps to follow the same IT specifications and the same sequencing of IS components disseminating the same messages. All Sumerian Homo Saps processed Thought structures containing the same protocols and procedures for operating in the safe havens, all under the control of their institutions.

The Sumerian priestly institution had developed their version of WritingITIS with a comprehensive set of IT features and specifications. They had also demonstrated the effectiveness of their IS applications, the primary keys to ensuring the dissemination of directed messages and conformity in Thought processing. Storytellers labelled the IT "cuneiform," a label that described the configuration of the ImageITIS products the Sumerian Homo Saps designed to inscribe their WritingITIS substrates.

Sumerian Homo Saps developed the WritingITIS substrates using Earthly components that storytellers labelled "clay" and "water." They mixed the clay and water in proportions that allowed them to form ITIS substrates they labelled "seals," "tablets" and "cylinders." The characteristics of this technology forced them to design ITIS peripherals that produced cuneiform or wedge-shaped ImageITIS products to capture Thought structures on clay substrates.

They then simulated EOS utilities that generated Earthly heat to eliminate the water from the clay/water mixtures of the soft ITIS substrates and to help bind the clay particles into solid, hardened products. The binding fixed the shape of the WritingITIS elements in the surfaces of the hardened substrates and the sequences of the elements into Thought structures to produce finished but fragile physical products. Individual Homo Saps could easily destroy the substrates.

Sargon chose to adopt the IT of the Sumerian version of WritingITIS without major changes. His ITISanian analysis indicated that developing a new IT would be counterproductive. The IT was flexible enough to allow him to make changes to the IS components and disseminate messages promoting his Akkadian Thought-processing controls.

He directed the Sumerian priestly institution to store the Sumerian WritingITIS products in facilities that storytellers labelled as "libraries." He designed the libraries as repositories—ITIS storage peripherals—that allowed Homo Saps access to the Sumerian WritingITIS products for reference and research.

His ITISanian purpose in isolating Sumerian WritingITIS products was to begin the development of new sets of WritingITIS products that re-sequenced Sumerian Thought structures to capture and disseminate Akkadian IS components. Since Akkadians used primarily OralITIS products to disseminate their control Thought structures, he directed scribes to adapt the features of Sumerian IS applications to capturing and inscribing the procedures and protocols of the Akkadian version of

OralITIS. He succeeded in making Akkadian OralITIS the common OralITIS tool for use throughout his empire with these directives.

Sargon's ITISanian Thought processing initiated ITIS processes whose outcomes were irreversible and inevitable. The ultimate outcome was the release of the individual Homo Sap Thought-processing imperative from the restrictions that the Sumerian institutions imposed to control the development and applications of ITIS tools and products.

The differences between the Akkadian and Sumerian OralITIS products, and their procedures and protocols for sequencing and capturing IS components, forced Sumerian Homo Saps to reactivate their Thought-processing facilities—facilities that the Sumerian priestly institutions had inhibited. The Sumerian Homo Saps had to capture the Akkadian OralITIS products and process them to sequence new Akkadian procedures and protocols for their HAP survival.

The processes forced individual Homo Saps to reprocess fixed Thought structures that directed their activities, remove Sumerian IS components, and then adapt the Thought structures to contain new Akkadian IS components disseminating Akkadian control messages. The reprocessing inevitably expanded individual Homo Sap memories. The free space outside fixed Thoughtbases automatically prompted the Homo Saps Thought-processing imperative to initiate new free-Thought sequencing and structuring.

The Akkadian Homo Saps developed their version of OralITIS with IS features radically different from those of the Sumerian version. Both Homo Sap groups implemented similar vocal technology, but developed different IT features, all the outcomes of COS and EOS utilities that stabilized HAP structures during the Earthly process of growing them into viable individual hominid animals. These utilities interacting with other utilities during the HAP maturing process fixed the embedded HAP vocal IT in different individuals at different levels of technological development.

Each hominid animal stabilized with unique sets of vocal IT features. Some stabilized with IT features that allowed them to implement their vocal technology to produce sounds that were louder than those of other individual hominid animals. Storytellers labelled this IT feature as "volume control." Some individual hominid animals stabilized with IT features that allowed them to implement their vocal technology to produce sounds that were different from those of other individual hominid animals activating the same IT features. Homo Saps labelled these IT features as "frequency and tone control." Some individual hominid animals stabilized without some IT features, leaving some individual hominid animals without the capability to produce certain sounds.

To replicate their OralITIS products in WritingITIS products, Sumerian Homo Saps reconfigured the ImageITIS products they originally developed for use in WritingITIS products they originally developed for the management and control of survival resources. They implemented an ITIS process that joined WritingITIS elements to represent standardized sounds and words they used to sequence their OralITIS products. Storytellers labelled the ITIS process "agglutination," a process user-friendly to the capturing of Sumerian OralITIS products in their WritingITIS products.

The Akkadian Homo Saps had developed a different set of HAP vocal IT features. They designed the ImageITIS elements of words to represent a sound feature of their HAP vocal IT. Storytellers labelled this feature as "syllabication," individual sounds "syllables," and the word elements of Sumerian OralITIS products as "mono-syllabic."

The agglutination process forced Sumerian Homo Saps to design new ImageITIS elements. Storytellers labelled them "diacritics." They connected the diacritics to the sequences of ImageITIS elements representing the compound words in WritingITIS products. The diacritics functioned as indicators of the purposes of the compound words. They helped scribes identify whether the ImageITIS products forming the word represented the real, sensible objects that the ImageITIS products captured or represented the abstractions that the IS component described.

Homo Sap groups which developed and implemented different IT features of their HAP vocal technology in the design of their versions of OralITIS had difficulties in capturing and processing OralITIS Thought structures that other Homo Saps designed. Different groups developed other IT utilities as well as the IT utilities producing agglutination and syllabication features that helped them complete the Thought structures defining IS components of their OralITIS products.

They invoked IT utilities whose routines allowed them to change the durations of sounds, change the frequencies of the sounds, the tones of the sounds, the volumes of the sounds. They invoked IT utilities whose routines directed them to implement specific movements of HAP structures as they delivered the OralITIS products. The IT utilities directed them to open and close hands in concert with specific sounds or sequences of sounds, to make fists, to expose open palms, to stamp feet, to stand still, to wave arms, to change expressions on the faces, show their teeth, and to keep their eyes open or closed. The combined outcomes of all the IT utilities that they could invoke or fail to invoke while delivering their OralITIS products completed the Thought structures that they sequenced to define the IS components of their OralITIS products.

The Sumerian priestly institutions implemented only the IT process of agglutination to specify compound word elements with which to sequence IS components describing abstractions. The process introduced complexities in their WritingITIS products. Only priestly Homo Saps were able to determine whether the IS components of the compound word elements referred to resource and business events or to abstractions that they defined only in individual Homo Sap memories. Homo Saps labelled the outcome of the IS component as "meaning."

Sargon sequenced Thought structures that described the effectiveness of the Sumerian resource production procedures and protocols. He adopted most of them without making changes to ensure the continued operation of the successful resource production and control system.

Sargon achieved his ITISanian objective of improving the user-friendliness of ITIS tools and his leadership objective of ensuring the continued dominance of Akkadian Homo Saps. The ITIS processes he initiated allowed him to promulgate Thought-processing controls but without the restrictions with which the Sumerian institutions inhibited individual Homo Saps from processing free-Thought in individual Homo Sap memories.

With this Thought-processing freedom, both Akkadian and Sumerian Homo Saps began demonstrating the ITIS productivity of individual Homo Sap Thought processing. They exercised their Thought processing facilities and began implementing the IT process of syllabication. They produced new words whose elements defined only the IT components of the ImageITIS products that they designed to inscribe the new words on WritingITIS substrates. They produced unique sequences of syllables that identified the precise messages of the

IS components in the Thought structures of WritingITIS products. They began standardizing the syllabic sounds and their ImageITIS representation as elements of the new words that they developed. They produced WritingITIS products containing syllabaries they defined as lists of syllables and the sounds that such ImageITIS elements represented.

Before Sargon and his two male Homo Sap offspring lost control of the Mesopotamian Empire to other Homo Sap groups, the Mesopotamian Homo Saps' continuous Thought processing to capture the new Akkadian messages conditioned the Mesopotamian Homo Saps to automatically adapt the Akkadian version of the ITIS tool of OralITIS to replace the less user-friendly Sumerian version. Instead of eliminating all applications of the two Sumerian ITIS tools, Sargon designated them as ITIS tools for the production of OralITIS and WritingITIS products containing Thought structures defining their godly entities and the procedures for transmitting and receiving godly Thought structures. Homo Saps labelled the WritingITIS products produced as "sacred" and the OralITIS products for transmitting and receiving godly Thought structures as "sacred rituals."

HAMMURABI

After Sargon and his two male Homo Sap offspring, other Homo Sap groups assumed control of the Mesopotamian safe havens. Homo Saps labelled these Homo Sap groups "Amorites, Gutians, Elamites, Hurrians, and Hittites." They entered the Mesopotamian safe havens from safe havens they had established on the western, eastern, northern, and northwestern fringes of the Earthly region that Mesopotamian Homo Saps controlled as an empire.

These Homo Sap groups initially focused on accessing the abundant survival resources that the Mesopotamian Homo Saps maintained. Their first actions focused on taking physical control of the Mesopotamian safe havens. They implemented the killing procedures of the HAP survival directive. They threatened Homo Saps with harm to the hominid physical structures of all those who opposed them.

Each Homo Sap group imposed Thought controls on the Homo Saps individuals who operated in the safe havens. With each set of new Thought controls, their HAP survival imperative constantly prompting them, Homo Saps activated their individual Thought-processing facilities to reprocess their individual survival Thought structures and incorporate the different IS components from the new Thought controls in their survival procedures. The constant need to activate individual Thought-processing facilities expanded individual Homo Sap memories and reinforced the Thought structures identifying the outcomes of Thought processing in individual Homo Sap memories as more powerful in ensuring individual physical survival than the outcomes of hominid animal killing procedures to establish the dominance of one animal species over another.

Most of the leaders of the Homo Sap groups which imposed their HAP controls on the Mesopotamian safe havens reprocessed their control Thought structures after they established their physical dominance over the safe havens. They re-sequenced the Thought structures to reach outcomes whose IS components identified the abundance of survival resources as outcomes of the application of ITIS tools that Mesopotamian Homo Saps designed to capture Thought structures specifying the most efficient procedures for managing survival resources. The leaders who fixed these Thought structures was the leader of the Amorite Homo Sap group. Homo Saps named him Hammurabi.

Storytellers described the Homo Sap Hammurabi as having the Homo Sap characteristic of cruelty, a characteristic that among all animal species only Homo Saps exhibit. He acquired consummate skills in applying the hominid animal procedures that Homo Saps re-sequenced with procedures for threatening and physically harming other Homo Saps. They labelled the procedures "torture."

Storytellers also defined his name as the concatenation of syllables from a sequence of words whose IS components identified him as a Thought processor of laws. This Thought processing and the ITIS events he initiated identify him as an ITISan.

Hammurabi imposed the four ITIS tools of OralITIS, ImageITIS, CalendarITIS, and WritingITIS on all Homo Saps who operated in the Mesopotamian safe havens and changed the Mesopotamian Empire into a Mesopotamian Homo Sap civilization. All four ITIS tools had their provenance in the original IT and IS designs of the Sumerian and Akkadian ITIS tools.

HAMMURABI'S THOUGHT-PROCESSING BASE

The Mesopotamian OS over which Hammurabi assumed control was one that had allowed ITIS events that prompted individual Homo Saps to continuously process and adapt to the procedures and messages in changing Thought structures. More and more individual Homo Saps, their HAP survival imperative dominant, had to acquire new ITIS skills in order to simply survive.

Hammurabi assumed control of Mesopotamian Homo Saps who had already stored and processed Sargon's and Urakagina's Thoughtbase. They had begun to free themselves from the ITIS and Thought-processing controls of their priestly institutions. This freedom automatically expanded both Thought processing and storage capacities in individual Homo Sap memories, the inevitable outcome of free Homo Sap Thought processing.

They sequenced and captured Thought structures in WritingITIS products whose IS components began to describe the OS of their cosmic and Earthly realities as independent systems whose operating utilities produced fixed ranges of physical outcomes; the precise outcomes were uncertain but within the fixed range. They stabilized the HAP structures of Homo Saps in unique configurations that made each individual Homo Sap unique. Storytellers labelled the relationship between the COS and EOS and the ultimate design of HAP structures as "astrology."

They sequenced Thought into structures whose IS components described the relationship between Homo Saps and the godly entities they chose as omnipotent, omniscient directors of their Earthly realities. They captured Thought structures in a storytelling application of WritingITIS that Homo Saps labelled as "epic." They labelled the first epic WritingITIS product as the Epic of Gilgamesh.

The storytelling Thought structures described the Sumerian Gilgamesh as part godly entity and part Earthly Homo Sap. The Thought structures described the physical and Thought-processing procedures he executed to access and retrieve the godly Thought structures containing the procedures whose godly outcomes assured him of the permanent viability of his HAP structure and neutralized the inexorable physical disintegration and reconstruction procedures through which the independent COS and EOS cycle all the physical structures of Earthly entities. Homo Saps labelled the outcome of these procedures as "immortality."

Though he retrieved the godly Thought structures, the Homo Sap Gilgamesh chose to ignore specific godly instructions and failed to achieve the outcome he set at the beginning of the events he initiated. He failed to capture the godly message in the IS component of the godly Thought structures. Storytellers described the message as an admonition to Homo Saps, assuring them that they would always fail in achieving outcomes they set for themselves without the godly instructions on the procedures for achieving the outcomes and without fully complying with the precise details of the godly procedures.

HAMMURABI BEGINS HIS ITIS DEVELOPMENTS

Hammurabi initialized his ITISanian Thought structures from the outcomes of those that Sargon and Urakagina disseminated. Like the Homo Saps Sargon and Urakagina, he captured the IS components disseminating ITISanian messages describing the power of ITIS tools in initiating and controlling the operation of Homo Sap safe havens.

Hammurabi expanded the IS components with Thought sequences containing the ITISanian message describing the most effective control strategies. He established the Akkadian versions of the ITIS tools of OralITIS and WritingITIS as the standard ITIS tools for all Homo Saps who operated in the Mesopotamian safe havens. He directed them to adopt these standardized ITIS tools in capturing and disseminating Thought structures throughout his Mesopotamian Homo Sap civilization. His ITISanian processing indicated that this was a critical ITIS event that would ensure all Homo Saps who chose to operate in Mesopotamian safe havens followed the same procedures and protocols in their Thought processing. Like Sargon had done, he left unchanged the applications of the Sumerian versions of both ITIS tools for capturing and disseminating godly, religious, and sacred messages.

These ITIS events helped him in unifying the different groups of Homo Saps who operated in his embryo Mesopotamian Homo Sap civilization. More and more Homo Saps began to acquire skills in the applications of the Akkadian versions of the ITIS tools and captured the same messages from IS components in the ITIS products. The outcome was an increasingly stable Mesopotamian OS and safe havens in which individual Homo Saps freely conditioned their Thought processing to execute the same procedures and comply with the same directives in the same IS components.

Hammurabi consolidated the ITISanian outcomes of his Thought control initiatives by implementing powerful ITISanian procedures. Storytellers labelled them "laws."

He produced WritingITIS products on which he sequenced identical sets of Thought structures containing IS components directing Mesopotamian Homo Saps to execute specific procedures in determining the outcomes of interactions between individual Homo Saps and the institutions supervising and maintaining the OS of his Mesopotamian Homo Sap civilization. He demonstrated his ITISanian skills by adapting his Thought structures to interface and amplify those that Urakagina initialized. Several generations of Mesopotamian Homo Saps had already processed Urakagina's Thought structures, captured and stored their procedures and IS components in individual Homo Sap memories, and automatically implemented the outcomes.

Hammurabi reprocessed Urakagina's Thought structures, designed the interface to transmit the original IS components and their messages, and structured the expansion sequences to specify the outcomes of interactions between individual Homo Saps and Mesopotamian institutions. By specifying the outcomes of these interactions he compelled all the aristocratic Homo Saps who supervised the Mesopotamian safe havens to implement the same outcomes for the same interactions in all safe havens.

Homo Saps labelled the combination of Thought structures, their embedded procedures and IS components, and their messages and outcomes as the "Laws of Hammurabi." With the completion of these ITIS events, Hammurabi modified the OS of the Mesopotamian safe havens, redirected the Thought processing of individual Homo Saps, and initiated changes in social structures whose outcomes had wider impacts on Mesopotamian Homo Saps than those of the Thought structures of the Homo Sap Urakagina.

Urakagina's Thought structures had specified sets of rules and procedures that storytellers labelled as "custom." The Homo Sap members of the aristocratic institutions assumed the primary responsibility for applying the rules and procedures and choosing the outcomes of the interactions of individual Homo Saps and institutions. The aristocratic

Homo Saps stored the Thought structures defining the customary rules and procedures in their individual Homo Sap memories and transmitted the Thought structures from one generation of aristocratic Homo Saps to another in OralITIS products.

The flexibility of Thought processing in individual Homo Sap memories allowed the aristocratic Homo Saps to modify the Thought structures containing the rules and procedures to specify outcomes they chose, often arbitrarily. The Thought-processing capabilities of different individual aristocratic Homo Saps and their skills in the production of ITIS products produced outcomes that differed from one safe haven to another.

Hammurabi fixed the rules and procedures and the outcomes of their applications. The IS components directed aristocratic Homo Saps to reach the same outcomes. By posting his WritingITIS products on monuments in Mesopotamian safe havens and along roads connecting them, Hammurabi compelled all Homo Saps operating in the Mesopotamian Empire to comply with his laws.

Hammurabi initialized a new set of IS components in the Thought structures of his laws. The IS components contained messages that focused on individual Homo Saps. The messages began to identify individual Homo Saps as Earthly entities whose freedoms to function outside the Thought controls of institutions were most critical to the survival of the Homo Sap species, both as hominid animals and as Thought processors. Hammurabi limited the individual Homo Sap freedom to kill other individual Homo Saps and specified harsh but precise penalties for those who damaged the HAP structures of other individual Homo Saps or who failed to comply with laws.

He also gave individual Homo Saps a new freedom. He gave them the freedom to operate private enterprises. He eliminated the ancient procedure that forced them to submit to the control procedures of institutions that for millennia had imposed their controls of business transactions and the benefits of business profits.

Hammurabi initialized another set of IS components containing messages that began the changes in the relationship of Homo Saps and the godly entities that Mesopotamian priestly institutions chose as omnipotent and omniscient directors and supervisors of events in the safe havens. He designed the messages to have both OS and ITISanian outcomes. The messages described the godly entities as approving his laws, but the ITISanian message declared that he, as an individual Homo Sap, sequenced Thought into structures specifying the laws in his individual Homo Sap memory without godly prompts and free of godly direction.

Of all the new messages that Hammurabi initialized, storytellers have consistently disseminated these two sets of messages in their OralITIS and WritingITIS products. One proclaims the power of individual Homo Saps when they free themselves of the controls of institutions. The other proclaims the independence and power of Homo Saps Thought processing in the privacy of individual Homo Sap memories, free of institutional or godly interference. Capturing and processing these messages, individual Homo Saps over many generations after that of Hammurabi have persistently repeated these messages in OralITIS and WritingITIS. As persistently, both institutions and godly entities have threatened individual Homo Saps with the destruction of their HAP structures for processing and disseminating the messages.

THOUGHT PROCESSING DEVELOPMENTS

Some Mesopotamian Homo Sap leaders, once they had established their Homo Sap groups as HAP controllers of the central safe havens with threats to their Homo Sap subjects' HAP structures, implemented features of their ITIS tools to reinforce and consolidate their controls. They focused on Thought-processing controls rather than HAP killing procedures. They directed their scribes to produce WritingITIS products whose IS components and messages promoted the dominance of only their Homo Sap groups.

They gathered all WritingITIS products that all previous dominant Homo Sap groups had produced. They stored the products in libraries. They designed ITIS systems with procedures that allowed them to control the storage and retrieval of WritingITIS products in the libraries. They imposed restrictions on the freedoms of individual Homo Saps, Mesopotamian, and others visiting the Mesopotamian safe havens, who wanted to access and capture or re-sequence and reprocess the Thought structures in any of the WritingITIS products.

Storytellers labelled these Homo Saps as "scholars," individual Homo Saps who had acquired skills with the ITIS tools and who dedicated their Earthly activities to processing Thought structures, sequencing new Thought structures, and re-sequencing older Thought structures to produce new outcomes. These Homo Sap scholars operated primarily in response to prompts from their Homo Sap Thought-processing imperative. Some were able to free themselves from the restrictions of existing Thoughtbases to access MEMORY and retrieve Thought sequences with which they expanded Thought structures, IS components, and messages in ITIS products that other Homo Saps produced with different versions of the ITIS tools of OralITIS and WritingITIS.

The ITISanian outcomes of this freedom to access and process the ITIS resources of the Mesopotamian libraries prompted Homo Saps to focus on the demands of their Homo Sap Thought-processing imperative. They increased the production of new ITIS products to sort and categorize the cascades of new Thought structures they were sequencing in individual Homo Sap memories.

Some Homo Sap scholars expanded Thought structures describing the physical design of their cosmic and Earthly environments. They expanded Thought structures identifying the procedures and utilities of the OS controlling their physical environments. They began to analyze the uncertainties of the OS outcomes on the physical survival of hominid and other animal species.

They expanded on Thought structures that defined physical tools which individual Homo Saps designed to enhance their Earthly physical capabilities. They improved the design of the tools that helped them reconfigure the physical environment to increase the production of survival resources or to increase the protection against damage to their HAP structures.

Some Homo Sap scholars sorted, reprocessed, and expanded Thought structures with IS components that disseminated messages identifying the relationships between Homo Saps and the godly entities they chose as the omnipotent, omniscient supervisors of Earthly physical events and directors of Homo Sap survival activities. They were unable to unequivocally identify real physical outcomes for the Thought structures defining the godly entities since the godly Thought structures had realities only in individual Homo Sap memories— thought environments without sensible restrictions.

The demands of their Homo Sap Thought-processing imperative compelled them to keep processing and reprocessing the godly Thought structures until they reached real, sensible outcomes. They produced and kept expanding networks of Thought structures with IS components whose messages identified godly entities infinite in number, all with differing attributes, Earthly powers, and relationships with each other and with individual Homo Saps or Homo Sap groups. They implemented the ITIS utility of naming, a powerful utility whose outcomes allowed individual Homo Saps to use the names as keys to retrieving from MEMORY the Thought structures specifying the godly characteristics and procedures for executing the Earthly powers of the godly entities.

The Homo Sap scholars failed to identify the true names of their godly entities. Automatically reacting to the demands of their Homo Sap Thought-processing imperative, they extended the godly Thought structures with sequences describing their failure as the outcomes of Thought inhibitions that the godly entities imposed on them. The IS components disseminated messages that the godly entities considered Homo Saps, as Earthbound entities, focused only on their HAP survival and incapable of executing godly procedures without annihilating themselves.

Some Homo Sap leaders attempted to destroy the WritingITIS products of groups they conquered. Their purpose was to re-sequence the Thought structures in the WritingITIS products, delete the IS components and messages praising the accomplishments of previous leaders, then reproduce the Thought structures with IS components and messages praising the new leaders for the same accomplishments. They were able to destroy physical ITIS products but they were never able to destroy their Thought structures, IS components, and messages. Individual Homo Saps always saved copies of the original products with their original messages or retained the original Thought structures in the privacy of their individual Homo Sap memories. Two such products were those whose Thought structures captured the messages in the storytelling Thought structures describing the Homo Sap Gilgamesh and the Homo Sap Hammurabi.

AFTER HAMMURABI

The Homo Sap groups, which established their HAP dominance of the Mesopotamian safe havens after the Amorite Homo Sap groups, all adopted the kernel of the OS Hammurabi had established. The kernel procedures defined an application interface that restricted any changes they made to modifications of peripheral utilities.

They left unchanged the IT features of the versions of the ITIS tools Hammurabi established to standardize the Thought-processing procedures. All the dominant Homo Sap groups modified primarily the IS components of the Thought-control structures, re-sequencing the Thought to disseminate messages proclaiming their control of the Mesopotamian safe havens. They all retained the original institutions to maintain the kernel procedures and utilities that had kept the Mesopotamian Empire functioning efficiently.

The stability of the OS allowed some leaders to direct their military institutions to extend the Earthly boundaries of the Mesopotamian Homo Sap civilization and establish their Thought controls on the OS of safe havens that other Homo Sap groups had established. They re-engineered the OS of newly-occupied safe havens and established system interfaces through which the core procedures and utilities of the Mesopotamian OS automatically executed to retain primary controls on all activities of the Homo Sap groups in the new safe havens.

Storytellers re-labelled the expanded Mesopotamian Homo Sap civilization with labels identifying the Homo Sap groups or leaders that initiated and consolidated the expansion. All exercised their HAP and Thought-processing controls over the Mesopotamian Homo Sap civilization from the original safe havens that the Sumerian and Akkadian Homo Saps established on the banks of the Euphrates and Tigris rivers.

With each expansion, the Mesopotamian civilization incorporated many other Homo Sap groups. Most had developed their own versions of the primary survival ITIS tools of OralITIS and ImageITIS. Some had also developed their own versions of WritingITIS. Their survival depended on their acquiring skills with the Mesopotamian ITIS tools, but they also maintained their own distinctive versions for use within their individual groups. Storytellers labelled these versions "vernacular."

The constant requirement to reprocess Thought structures into one or another version had an inevitable outcome. Individual Homo Saps expanded their Thought processing and began sequencing and re-sequencing new and old Thought structures in the free space of individual

Homo Sap memories. They produced new and modified outcomes and isolated messages stripped of their institutional context.

Storytellers began to separate the messages from their original storytelling Thought structures and capture them in new sets of Thought structures. The new Thought structures amplified and explained the messages. The outcomes unequivocally began to describe individual Homo Saps as unique Earthly entities with independent Thought-processing facilities in individual Homo Sap memories. Each individual Homo Sap had the Thought-processing capability to sequence Thought into structures in the privacy of individual Homo Sap memories with procedures that each was free to execute and implement outcomes that each chose to implement.

The Homo Sap groups which produced the clearest messages were those which operated in safe havens on the periphery of the Mesopotamian Homo Sap civilization. Their distance from the central Thought-processing controls diminished the power of the institutions to monitor and ensure the impact of the controls. The Mesopotamian Homo Saps who supervised operations in these remote safe havens focused more on ensuring the indigenous Homo Sap groups paid their taxes and tributes and the profits flowed to the core Mesopotamian institutions. These distant safe havens had few of the Mesopotamian ImageITIS products that functioned in the core Mesopotamian safe havens to continuously disseminate their embedded messages proclaiming the omnipotence of Mesopotamian institutions. Without these constant Thought-processing controls, they were free to sequence and process new Thought structures.

Development of AlphabetITIS

The number of Homo Saps who acquired WritingITIS skills as a necessary survival tool grew rapidly in the 20 generations of Homo Saps Thought processing after Hammurabi. In both the established, institutionally-controlled civilizations of Ancient Egypt and Mesopotamia, the need to ensure HAP survival in operating environments with a complex of ITIS tools, protocols, and procedures prompted many individual Homo Saps to begin to develop an IT capable of capturing IS components user-friendly to a broad range of Homo Sap groups.

Different Homo Sap groups had been developing ITs that allowed them to design applications of WritingITIS that emulated and captured the IS components of their versions of OralITIS. The Sumerian Homo Saps developed the IT process of agglutination to capture their OralITIS Thought structures in their WritingITIS products. The Akkadian Homo Saps developed syllabication. These IT processes expanded the applications of the cuneiform version of WritingITIS to capture and disseminate Thought structures and messages they would traditionally disseminate in their OralITIS products. The outcomes were successful in making the cuneiform version of WritingITIS more user-friendly, but only to those who had skills in the procedures and protocols of Sumerian/Akkadian versions of OralITIS.

These user-friendly adaptations of Akkadian WritingITIS had little effect on the broad range of Homo Sap groups which operated in the Mesopotamian Homo Sap civilization. The adaptations captured only the unique sounds that Akkadians produced with their HAP vocal IT when delivering their OralITIS products to other Homo Saps. Most of these subject Homo Sap groups had their own versions of OralITIS and WritingITIS. Their HAP aural IT responded to different combinations of sounds. They still had to acquire skills in ITIS tools of their Akkadian administrators to operate successfully in the Mesopotamian safe havens.

Another ITIS impediment to the user-friendliness of Akkadian WritingITIS were the number of ImageITIS products they developed as WritingITIS IT elements. To acquire skills with the Akkadian ITIS products, individual Homo Saps had to store hundreds of ImageITIS products in their Homo Sap memories together with Thoughtbases that specified the application and meaning of each ImageITIS product. The same was true of the hieroglyphic and hieratic versions of Ancient Egyptian WritingITIS.

The successful development of the IT processes of agglutination and syllabication led some individual Homo Saps to begin re-sequencing the agglutination and syllabication Thought structures. The re-sequenced Thought structures described HAP vocal IT as capable of producing single, much simpler sounds than the outcomes of agglutination and syllabication IT processes. The IS component contained the message that all Homo Sap groups, though they had developed a wide range of OralITIS specifications, were capable of activating their HAP vocal IT to reproduce the same set of single sounds.

This outcome prompted Homo Saps of many groups to begin the

development of a new ITIS tool capable of making WritingITIS more user-friendly without institutional involvement. Individual Homo Saps were to increase the speed of the development with each succeeding generation.

Storytellers gave the label "Alphabet" to the new ITIS tool.

The Ancient Egyptian Homo Saps reintroduced 24 simple ImageITIS products as their version of AlphabetITIS during the reign of a god-king that storytellers named "Ramses," but the event failed to prompt individual Homo Saps to adopt the ITIS tool. The 24 ImageITIS products were hieroglyphic symbols and still evoked the Ancient Egyptian association of sacredness and institutional WritingITIS Thought-processing controls.

A Homo Sap group which operated in the Mesopotamian Homo Saps civilization and developed versions of OralITIS and WritingITIS that storytellers labelled "Aramaic" had more success. They designed a form of AlphabetITIS that made Aramaic the most widely used OralITIS and WritingITIS tools for business and inter-group transactions in Mesopotamia. The ITIS products still had to conform to Mesopotamian institutional protocols and procedures and were user-friendly only in Mesopotamia.

The Homo Sap group which produced the most influential version of AlphabetITIS operated in safe havens bordering a sea on the western fringes of the Mesopotamian Homo Sap civilization. Storytellers labelled the group "Phoenician" and the sea "Mediterranean."

The Phonecian Homo Saps operated primarily as sea-faring traders. Their trading partners and customers operated safe havens along the shores of the Mediterranean far outside the institutional controls of the Mesopotamian and Ancient Egyptian Homo Sap civilizations. Most developed their versions of OS and ITIS tools, protocols and communication procedures different from those the Phonecian Homo Saps implemented. These differences made communications difficult and trading agreements complex.

To simplify their business transactions with customers whose ITIS tools were different from their own, the Phonecian Homo Saps developed a system of 22 ImageITIS products that represented only the sounds that hominid animals were capable of producing with their vocal IT. They designed simple configurations for the 22 ImageITIS products, each user-friendly to reproduce on a WritingITIS substrate, each without any connection to hieroglyphic or cuneiform ImageITIS products or to the sacerdotal, institutional, or god-king Thought structures, with fixed protocols and communication procedures and messages embedded in the established WritingITIS systems.

Homo Saps who began developing the new IT and demonstrating the ITIS applications of AlphabetITIS embedded an IS component in the Thought structure specifying the IT. The IS component captured and disseminated a powerful message for individual Homo Saps. The message described the IT as most user-friendly to individual Homo Saps as a tool to capture and isolate in WritingITIS products Thought structures that they individually sequenced into structures in the privacy of individual Homo Sap memories.

Isolating Thought structures in WritingITIS products was a critical ITIS event. The capability allowed individual Homo Saps to return to the original Thought structures and reprocess and expand them without interference from other Thought structures in their individual Homo Sap memories.

Implementing the IT of AlphabetITIS, individual Homo Saps began to acquire WritingITIS skills without submitting to the ITIS and Thought controls that the institutions imposed on the Homo Saps they chose to train as scribes. Eliminating institutional ImageITIS products and the fixed Thought structures they represented from WritingITIS products, Homo Saps also eliminated the embedded IS components and messages continuously reinforcing Thought structures that assigned institutions absolute control over Thought processing in individual Homo Sap memories and the safe haven OS.

AlphabetITIS' user-friendliness allowed Homo Sap groups to capture in WritingITIS products the IT and IS features of their vernacular versions of OralITIS. Homo Saps described OralITIS products as vernacular when the IT and IS features differed from those that institutions specified for OralITIS products in the safe havens under their control.

They designed IS components whose messages focused on godly entities and, by association, the aristocratic and priestly institutions which claimed the godly authority to control the Thought processing and activities of individual Homo Saps. The messages identified godly entities as the abstract outcomes of Thought structures that Homo Saps had the individual freedom to ignore in directing individual Homo Sap activities in Earthly environments.

In each generation the outcome of the AlphabetITIS development allowed individual Homo Saps to begin acquiring the facility to sequence Thought into structures identifying relationships between godly entities, aristocratic and priestly institutions, and individual Homo Saps. The relationships were different from those that their institutions specified and imposed on Homo Saps since the Sumerian and Ancient Egyptian Homo Saps established their safe havens.

THE IMPACT OF ALPHABETITIS IN THE INSTITUTIONAL ENVIRONMENTS

The power of generations of institutional Thought-processing controls overwhelmed the power of the new ITIS tool of AlphabetITIS in the two dominant Homo Sap civilizations. Many generations of Homo Saps had established internetworks of Thought structures with complex interacting procedures that ensured the compliance of individual Homo Saps with institutional controls in the safe havens and in the Earthly and Thought-processing centers of the civilizations. The stability of the entire structure of the safe havens depended on all Homo Saps executing the procedures that produced compliant outcomes. The complexity of the internetwork of control procedures allowed individual Homo Saps to design individual procedures with outcomes compatible with the requirements of the institutional control procedures.

Their HAP survival imperative controlling their Thought processing, both individual Homo Saps who chose to operate in the central safe havens and those who joined institutions, inhibited themselves from processing any Thought structure whose outcome threatened the orderly operation of the safe havens. Confining their Thought processing to the internetworks of control Thought structures and their fixed outcomes, they automatically reacted to oppose any changes in procedures that guaranteed their physical safety and the availability of abundant survival resources. The power of their HAP survival imperative overwhelmed that of their Homo Sap Thought-processing imperative and they suppressed the development of AlphabetITIS in the central safe havens.

They left the development of the AlphabetITIS to Homo Saps who operated in safe havens on the peripheries of the Earthly region that the Mesopotamian institutions controlled, safe havens remote from the OralITIS, ImageITIS, and WritingITIS products that generations of Homo Saps had designed with IS components containing messages proclaiming only their institutions as the guarantors of their survival resources.

Storytellers labelled the Homo Saps who developed the most powerful version of AlphabetITIS as "Ancient Greeks." The Ancient Greek Homo Saps operated just outside the eastern boundary of the Mesopotamian Homo Sap civilization. Their development as a Homo Sap group free of institutional controls allowed individuals to develop and implement the ITIS tool in just ten generations. Storytellers labelled the single sounds and the elements of AlphabetITIS representing them as "consonants" and "vowels."

The Ancient Greek Homo Saps inherited and reprocessed many new Thought structures that other Homo Saps sequenced after the development of the first versions of AlphabetITIS. Homo Saps in Ancient Egypt and Mesopotamian had already initialized new Thought structures. They had begun reconfiguring Thoughtbases and Thought structures to which institutions had confined their Thought processing for hundreds of generations.

9:

ITIS IN ANCIENT GREECE WITH COMMENTS ABOUT ITIS IN ANCIENT INDIA AND ANCIENT CHINA

Dear Jim:

Three individual Homo Saps, each belonging to Homo Sap groups operating in different Earthly regions, initiated beginnings within four Homo Sap generations of each other, 125 Homo Sap generations after that of the Ancient Egyptian god-king Narmer. Each demonstrated free-Thought-processing capabilities and disseminated ITISan skills. Only one would initialize Thought structures that others would develop into Thoughtbases free of institutional controls and whose Thought structures defined the OS of a new Homo Sap civilization that storytellers label "Western." The other two had to develop their Thought structures in established institutional environments.

Storytellers gave the name "Thales" to the first individual Homo Sap. Thales operated in the safe haven of Miletus on the western edge of the Mesopotamian Homo Sap civilization that shared a border with the Earthly region that storytellers labelled "Ancient Greece." Storytellers gave the name "Buddha" to the second individual Homo Sap. Buddha operated in the Earthly region that storytellers labelled Ancient India. Storytellers gave the name "Confucius" to the third individual Homo Sap. Confucius operated in the Earthly region that storytellers labelled "Ancient China." The Homo Saps of each group implemented different versions of the first four ITIS tools of OralITIS, ImageITIS, CalendarITIS, and WritingITIS.

Each Homo Sap initialized unique sets of Thought structures with outcomes and IS components disseminating messages that directed the Thought processing in the individual Homo Sap memories of their Homo Sap groups. All three were highly successful ITISans promoting the power of individual Homo Sap Thought processing.

Only the Homo Saps of Ancient Greece developed and implemented a user-friendly AlphabetITIS. The speed with which they adapted their version of AlphabetITIS to the development and production of WritingITIS products is a powerful testament to the power of the individual Homo Sap Thought-processing imperative. Individual Homo Saps immediately and fearlessly respond to their Thought-processing imperative to adopt any ITIS tool that frees their Thought processing from institutional restrictions. Ancient Greek AlphabetITIS was such a tool.

ITIS Restrictions in Ancient India and Ancient China

The Homo Saps of Ancient India and Ancient China operated in Homo Sap civilizations in which established institutions controlled the OS and the stabilizing ITIS tools of OralITIS, ImageITIS, CalendarITIS, and WritingITIS and the IS components and messages they captured and disseminated. Both Homo Saps Buddha and Confucius had to sequence their Thought structures with IS components disseminating messages that conformed with those of their controlling institutions. Both proclaimed the freedom of individual Homo Saps to sequence and process Thought structures in the privacy of their individual Homo Sap memories and to control their individual Earthly activities, but they also alerted individual Homo Saps to adjust their Earthly activities to conform to the requirements of their controlling institutions.

Both Homo Saps Buddha and Confucius had to capture their Thought structures in ITIS products whose specifications the institutions designed and controlled. The institutional IS components and messages embedded in the IT and Thought structures of their institutional ITIS products obscured the fundamental messages in the Buddhist and Confucian Thought structures. The institutions remained supreme controllers in both the Ancient Indian and Ancient Chinese Homo Sap civilizations. The institutions restricted and controlled ITIS and

Thought-processing developments for most of the next 125 Homo Sap generations after those of the Homo Saps Buddha and Confucius.

THE COSMIC GOD THOTH AND FREEDOM FROM INSTITUTIONS

One Ancient Greek Homo Sap—storytellers gave him the name "Socrates"—identified the cosmic god Thoth as the inventor of ITIS tools capable of freeing individual Homo Saps' Thought processing from confinement in moribund institutional control Thoughtbases. He hid the message in the context of a set of Thought structures that disseminated other more obvious messages. These messages identified the institutions' historical control of Thought processing and the dangers to institutions of uncontrolled WritingITIS products in individual Homo Saps' memory that Homo Saps captured with OralITIS and ImageITIS products.

A priestly Ancient Egyptian Homo Sap to whom storytellers gave the name "Manetho," confirmed the hidden message. The Homo Sap Manetho was an Earthly representative of the cosmic god Thoth. He disseminated the cosmic god Thoth's Thought structures and produced a large number of WritingITIS products, following his godly patron. The Homo Sap Manetho, obeying the directives of his HAP survival imperative, generally kept his Thought processing confined to the control Thoughtbases of Ancient Egyptian institutions.

But as a follower of Thoth, he had to clarify the Homo Sap Socrates' message.

The Homo Sap Manetho sequenced a set of Thought structures that explained the ITIS advantage of the Ancient Greek Homo Saps. The messages in the IS components described the Ancient Greek Homo Saps as fortunate to have developed their Thought processing with ITIS tools free of institutional Thought-processing controls.

Storytellers described—more precisely, speculated on—the causes of the absence of all-controlling institutions in the Ancient Greek Earthly environment. Their speculations focused on the outcomes of Earthly events that the EOS initiated during more than 50 Homo Sap generations before that of the Homo Sap Thales.

Storytellers identified the primary Earthly event as a volcanic eruption. The outcomes decimated Homo Sap groups that storytellers label as "Minoan, Ionian, and Mycenaean," virtually destroying their controlling institutions and safe havens. The EOS utilities initiated a catastrophic eruption of the volcano that vented through the island of Thera, close to the center of the Ancient Greek Earthly region. The force of the eruption devastated the region and destroyed most of the safe havens in which the Minoan, Ionian, and Mycenaean Homo Sap groups and their institutions operated. After the last of the Earthly events, other Homo Sap groups migrated to the Ancient Greek Earthly region and joined with the Homo Sap survivors of the Minoan, Ionian, and Mycenaean Homo Sap groups to form the Ancient Greek Homo Sap group.

The Ancient Greek Homo Saps, free of their institutions' absolute controls on Thought processing and ITIS developments, initialized their version of AlphabetITIS with the 22 simple ImageITIS products of the Phoenician Homo Sap communications protocol. They began to implement the 22 simple ImageITIS products as the IT for the production of OralITIS products and for capturing their Thought structures and IS components in WritingITIS products. The IS components of the ImageITIS products disseminated one message: Individual Homo Saps in any Homo Sap groups were free to implement the ImageITIS products in any WritingITIS applications they chose to design.

The Ancient Greek Homo Saps' development of their version of the ITIS tool of AlphabetITIS began the release of individual Homo Sap Thought processing from the structures and constrictions of godly and institutional Thoughtbases. The user-friendliness of the new Ancient Greek version of WritingITIS prompted individual Homo Saps to begin capturing their individual OralITIS products and isolating the Thought structures that they sequenced in the privacy and free space of their individual Homo Sap memories in WritingITIS products. With the new user-friendly ITIS tools, they began to produce a wide range of WritingITIS products and initialized the Thought structure whose IS component disseminated the message describing the power of user-friendly ITIS tools in freeing Thought processing in individual Homo Sap memories.

With their development of applications for their two new ITIS tools, the Ancient Greek Homo Saps made real the godly and institutional fears that the Ancient Egyptian god-king Narmer feared: user-friendly ITIS tools allowed individual Homo Saps to develop the freeing power of individual Thought processing, expand individual

Homo Sap memories, and ignore godly and institutional Thought-processing directives.

They could access the infinite Thought store of MEMORY, a store that only gods had access to. With this access, individual Homo Saps could begin to retrieve free-Thought from MEMORY. They could sequence the free-Thought into structures and specify procedures that each individual Homo Sap devised to achieve Earthly outcomes that they individually chose for their HAP operations in their chosen Earthly environments, ignoring the directives of their gods, god-kings, and institutions.

THE ANCIENT GREEK HOMO SAP CIVILIZATION

The Ancient Greek Homo Sap groups established themselves as a unique civilization with their implementation of their version of user-friendly AlphabetITIS.

Two Ancient Greek Homo Saps produced WritingITIS products that identified them as two of the first to demonstrate the power of their user-friendly AlphabetITIS. Storytellers gave them the names "Homer" and "Hesiod." Both demonstrated the Thought processing freedom their new user-friendly version of WritingITIS gave them to compile Thought structures and embed them with IS components and messages describing the reality and unreality of Ancient Greek Homo Saps' interactions with godly and aristocratic institutions.

Both the Homo Saps Homer and Hesiod—their individual Homo Sap Thought-processing imperative compelling them—undertook the task of retrieving Thought structures they stored in their individual Homo Sap memories and captured the Thought structures in WritingITIS products with their unique version of WritingITIS. They compiled their Thought structures using protocols for an ITIS application that storytellers labelled "Poetry," an original application of the Ancient Greek version of OralITIS for sequencing and disseminating messages of past events. The design of the Poetry protocols allowed them to sequence IS components expressing powerful hominid animal and Homo Sap emotions. They took full advantage of the simple ImageITIS products of AlphabetITIS—storytellers labelled them "open source freeware"—and their freedom from the godly and institutional messages embedded in and limiting the message flexibility of the proprietary, unchangeable ImageITIS products that the Ancient

Egyptian and Mesopotamian institutions specified and implemented in their WritingITIS products.

HESIOD

The Homo Sap Hesiod produced several WritingITIS products using applications of the Poetry protocols that focused on the impact on Homo Sap Thought processing. In one, he designed and disseminated messages in Thought structures describing the stifling dominance of old Ancient Greek aristocratic institutions in the control of the older Ancient Greek safe havens. In another, he detailed the operation of a farm. The Poetry protocols he chose disseminated messages that described the impact of the work as damaging to HAP structures and as inhibiting and depressing on Homo Sap Thought-processing control.

The Homo Sap Hesiod's most powerful WritingITIS product outlined a godly provenance for the new Ancient Greek Homo Sap civilization. He compiled his godly Thought structures from the OralITIS products of a group of Homo Saps who operated as professional tellers of tales and the official disseminators of messages in ITIS products before the Ancient Greek Homo Saps began producing WritingITIS products. Storytellers labelled them "Bards." Each of the forerunner Homo Sap groups of the Ancient Greek Homo Saps—the Minoan, Ionian, and Mycenaean groups—supported their own bards who kept disseminating the godly Thought structures, often modifying them from generation to generation to satisfy the Thought-processing requirements of each generation.

The bards compiled their OralITIS products—as did many Ancient Greek Homo Saps—from Ancient Egyptian Thought structures. They established a godly provenance for the new Ancient Greek Homo Sap civilization. They modified Ancient Egyptian Thought structures, re-sequencing IS components to give their messages an Ancient Greek caste. They re-sequenced the Thought structures that Ancient Egyptian priestly Homo Saps processed to describe their founding godly group of the cosmic gods Shu and Tefnut, Geb and Nut, Ra and Thoth, and the Earthbound godly siblings Osiris, Isis, Seth, Nephthys, and Horus. The re-sequenced Thought structures identified a replacement pantheon of Ancient Greek gods, all Earthbound.

The ImageITIS products the Ancient Egyptian priests designed to give their gods physical form described them as a mixture of species—a cow head on a hominid animal body, a falcon, a baboon, an Ibis head on a hominid animal body, an Ibis alone. Their gods were magical, unreal entities. The Ancient Greek bards gave their gods hominid animal forms. They functioned as Earthbound gods, but with immortal, indestructible HAP structures and Homo Sap Thought-processing capabilities far superior to their client Ancient Greek Homo Saps. All exhibited characteristics similar to those of the Ancient Egyptian gods but as individual Homo Saps.

The Homo Sap Hesiod captured the godly Thought structures of the old bards. They described two groups of gods. Storytellers labelled them "Titans" and "Olympians." The old bards described the Olympians as the godly offspring of the Titans.

The Titans began with Gaea, a female Titan, an outcome of THOUGHT out of Chaos—as was the supreme but sexless Ancient Egyptian god Atum. Gaea cloned her son Uranus, seduced him into serving her sexual requirements, and together they produced Chronos and other Titans. Gaea then forced Chronos to castrate his father Uranus and made Chronos the chief Titan. Chronos seduced his aunt Rhea and they began producing offspring, which Chronos immediately swallowed in fear of what they would do to him when they grew up.

Rhea saved one son, Zeus, by tricking Chronos into swallowing a stone instead of the newly-born Zeus. Zeus grew up, defeated Chronos in a revolt against the Titans. He banished Chronos and some of the other Titans to Tartarus, the Ancient Greek equivalent of Ancient Egyptian Atumic cosmic limbo, but Chronos escaped to operate in Italy as the god Saturn.

Zeus played a game of lots with his Titan brothers Poseidon and Hades—the storytelling Thought structures re-sequencing and modifying the Thought structures describing the dice game between the cosmic gods Ra and Thoth. The winner would assume the position of chief god. Zeus won and made himself chief god, as did Ra.

Zeus exhibited Ra's characteristics. He acted mostly in anger at his frustration over his inability to enforce an absolute godly authority on the other gods and the Homo Saps. Like Ra, he vented his anger by throwing lightning bolts and initiating Earthly disasters in futile attempts to make Homo Saps and other gods obey his directives and express respectful homage to him as chief god.

Aligning Zeus on Ra, the bards disseminated the message

proclaiming Homo Saps' freedom from godly control, for Ra had had to accept his godly limitations in controlling Homo Saps on Earth.

Zeus seduced Hera, his sister, and other female gods to produce the Olympians. He seduced Maia to produce the Olympian Hermes, the Ancient Greek facsimile of the Ancient Egyptian cosmic god Thoth. Aligning Thoth on Hermes was to have powerful impacts on Homo Saps' future development of ITIS products and their mastery of their own fate.

HOMER

The Homo Sap Homer produced two WritingITIS products in which he sequenced Thought structures with IS components outlining a Homo Sap provenance for the new Ancient Greek civilization. The IS components disseminated many messages. Some paralleled those the Homo Sap Hesiod disseminated about the old, defunct aristocratic institutions but with positive outcomes. Homo Sap leaders displayed Thought-processing capabilities and ideal Homo Sap characteristics in their interactions with other leaders and their followers. Individual Homo Saps freely questioned their leaders but remained loyal and obeyed directives.

The Homo Sap Homer embedded messages within messages, each identifying individual Homo Sap actions and their potential outcomes. One described a war and the hominid animal killing sprees as the outcomes of the unbridled expression of a female Homo Sap's hominid animal sexual imperative. Storytellers gave her the name "Helen." Another described the honor a female Homo Sap earns when she chooses to express her hominid animal sexual imperative with only one male Homo Sap. Storytellers gave her the name "Penelope" and her sole sexual partner "Odysseus."

A critical message identified new connections and interactions between Homo Saps and their gods. He made clear the older bards' Thought structures describing individual Homo Saps as processing thought structures with IS components disseminating messages describing new relationships with their gods. These new individual Homo Sap Thought-processing developments paralleled the Phonecian Homo Saps' beginning development and dissemination of their simple version of AlphabetITIS 20 Homo Sap generations before the events that the Homo Sap Homer captured in his two WritingITIS productions.

The Homo Sap Homer's messages clearly proclaimed Homo Saps' independence and freedom from the absolute control of gods. They were free to sequence and process Thought structures with outcomes they chose without the sanctions of their gods.

The Thought structures described the impotence of the gods in their attempts to regain control of individual Homo Saps who ignored the limits that the gods imposed on their freedoms to take independent Earthly actions, often in defiance of godly directives. Storytellers labelled these apostate individual Ancient Greek Homo Saps "heroes." Their actions and outcomes demonstrated the new Earthly reality of the relationship between Homo Saps and gods.

The new Ancient Greek Homo Saps, free to capture the messages in the Homo Saps Hesiod and Homer's WritingITIS products, processed them continuously in the 20 generations of Ancient Greek Homo Saps after the Homo Saps Homer and Hesiod compiled their WritingITIS products. They re-sequenced the initializing Thought structures, expanding IS components and messages, identifying outcomes free of godly and institutional obfuscations.

The Homo Sap Homer's WritingITIS products disseminated sets of Thought structures that provided the Ancient Greek Homo Saps a Thoughtbase from which to design an OS and found their own civilization five Homo Sap generations before that of the Homo Sap Thales. Ancient Greek Homo Saps directed their offspring to capture and store the Thoughtbase in their individual Homo Sap memories and process the Thought structures.

Storytellers identify the Homo Sap Thales as the first to sequence unique Thought structures with IS components disseminating messages focusing on the OS and reality of the Earthly environment free of a godly context. He initiated 15 generations of individual Thought processing during which more and more individual Homo Saps acquired WritingITIS skills, producing many versions of the Homo Saps Homer and Hesiod's IS components and embedded messages. They began to develop the open-source freeware of the user-friendly Ancient Greek AlphabetITIS with applications that isolated and expanded their own and other Homo Saps' Thought structures to design new Thoughtbases. They began to identify and store in a separate Thoughtbase all the proprietary controls that the institutions—godly, priestly, and aristocratic—had imposed on individual Homo Saps' freedoms to develop and implement applications for their ITIS tools.

The Outcomes of Ancient Greek Thought Processing Freedom and ITIS Developments

A Homo Sap generation after that of the Homo Sap Thales, individual Ancient Greek Homo Saps who operated in the safe haven storytellers labelled "Athens," executed the first successful revolution against absolute aristocratic institutional controls. They sequenced a set of procedures with which they formed the kernel of an OS whose outcome storytellers labelled "democracy." Only another generation later their democratic Thought structures prompted the Ancient Greek Homo Sap leader to whom storytellers gave the name "Pericles" to implement the first prototype of a governing institution dedicated to implementing a democratic OS.

The Ancient Greek Homo Saps who claimed membership in the traditional Aristocratic institutions never stopped sequencing and processing Thought structures whose IS components and outcomes continuously disseminated the message that only they had the prerogative to function as Homo Sap leaders. Their guiding Thought structures were those that described the successes, stability and longevity of the institutionally controlled civilizations of the Ancient Egyptian and Mesopotamian Homo Saps.

Unable to stop the democratic Thought processing of the Athenian Homo Saps, the aristocratic Homo Saps had to operate in accordance with the procedures of the democratic OS. By executing the procedures, they were able to assume partial control of the democratic OS. They modified procedures to produce outcomes that gave only aristocratic members of the democratic institutions privileges to access and execute the kernel operating utilities. They sequenced procedures that specified levels of democratic procedures to which they restricted other members of the democratic assemblies.

The aristocrats were unable to re-establish traditional aristocratic controls. Their primary impediments were the user-friendly ITIS tools. They were unable to control ITIS developments. Individual Ancient Greek Homo Saps constantly challenged them with the outcomes of their freedom to acquire and hone ITIS and Thought-processing skills with the user-friendly Ancient Greek version of WritingITIS.

For the next five generations, other Ancient Greek Homo Saps, particularly those storytellers labelled "Spartans," most of whom operated in safe havens under the control of aristocratic institutions, executed their HAP survival procedures to destroy the Athenian Homo Saps. They feared Athenian democratic successes would undermine the institutional OS of their safe havens. These traditional institutional challenges failed to stop the free-Thought-processing that individual Ancient Greek Homo Saps had begun to identify as their right.

WRITINGITIS APPLICATIONS AND INDIVIDUAL HOMO SAP MEMORY

One Ancient Greek Homo Sap to whom storytellers gave the name "Simonides" produced a WritingITIS product that began the development of the new WritingITIS application that storytellers labelled as "Prose." The Homo Sap Simonides was a prolific producer of ITIS products that he designed to entertain clients who paid him a fee for his ITIS production. His most influential WritingITIS product disseminated messages identifying the power of individual Homo Sap memory and the gruesome fate of aristocratic Homo Saps who arbitrarily chose to dishonor contracts.

The storytelling Thought structures describe the commitment an aristocratic Homo Sap made with the Homo Sap Simonides in payment for his entertainment services at a banquet the aristocratic Homo Sap gave for other aristocratic Homo Saps. The aristocrat instructed Simonides to design and disseminate at the banquet an OralITIS product with IS components lauding the host aristocratic Homo Sap's aristocratic progenitors, their godly connections, and their less-than-truthful Earthly achievements.

After the Homo Sap Simonides delivered his OralITIS product at the banquet in fulfillment of his contractual commitment, the host aristocrat complained that Simonides had failed to meet the intent of the aristocrat's instructions. The host aristocrat claimed that Simonides had failed to give his progenitors and their godly connections the praise they deserved and offered to pay only half of Simonides' agreed fee.

The host aristocrat's perfidy prompted the Ancient Greek Olympian gods to intervene in the dispute. Two Olympian godly entities to which storytellers gave the names "Castor and Pollux," the offspring of the supreme Ancient Greek god Zeus after his sexual conjugation with a female Homo Sap, arrived at the entrance to the banquet hall. They instructed a servant of the aristocratic host Homo Sap to bring Simonides out of the banquet hall to meet with them.

When Simonides was clear of the banquet hall structure, the supreme Ancient Greek god Zeus executed an EOS utility that initiated a "smart" earthquake by destroying only the parts of the banquet hall immediately adjacent to the banquet table. The outcome of the smart earthquake destroyed beyond recognition the HAP structures of the host aristocrat and his aristocratic guests at their places around the banquet table. The relatives of the aristocratic Homo Saps were unable to identify the remains of their kin.

Simonides offered to identify the HAP remains of each aristocratic Homo Sap for a further fee. He promptly identified the remains, demonstrating his ITIS methodology for storing Thought structures in individual Homo Sap memories and instantly retrieving them on command. The surviving Homo Saps at the banquet began to disseminate Thought structures describing the Homo Sap Simonides' feat.

Other individual Homo Saps' desire to acquire the same ITIS skills prompted the Homo Sap Simonides to sequence Thought structures defining his procedures for designing an ITIS methodology that allowed any Homo Sap to retrieve Thought structures from Homo Sap memories instantly, without error. For generations of Homo Saps after that of Simonides, individual Homo Saps re-sequenced and expanded the Homo Sap Simonides' Thought structures into systems that allowed individual Homo Saps to develop individual Homo Sap memories as an ITIS peripheral for accessing the infinite Thought store of MEMORY.

The ITIS Limitations of Poetry Communication Protocols

Ancient Greek professional tellers of tales, the bards, devised the protocols of poetry as a set of rules with which to sequence and store Thought structures permanently in their Homo Sap memories for transmission to other Homo Saps as OralITIS products. Poetry protocols allowed the bards to sequence IS components with messages that they focused on entertaining the Homo Saps who paid them to transmit the Thought structures. They designed the protocols with a flexibility that allowed them to re-sequence and focus the messages to satisfy the specific requirements of their Homo Sap audiences.

The bards constantly redesigned the IS components and messages to make unreal events real, always attempting to meet the demands of paying clients. The poetry protocols allowed them to immediately re-sequence a set of fixed Thought structures—storytellers labelled them "boiler-plate"—to produce or modify outcomes that their Homo Sap audiences wanted to capture and process in their individual Homo Sap memories. The same storytelling Thought structures that they disseminated to different Homo Sap audiences reached different outcomes, each outcome focused on the prejudices and hominid animal desires of their Homo Sap audiences.

Most individual Ancient Greek Homo Saps lacked the freedom from immediate HAP survival responsibilities to apprentice themselves to master Ancient Greek bards for the many primary Sun cycles they needed to train their individual Homo Sap memories to capture and store for rapid retrieval and transmission the standard storytelling Thought structures that the master bards designed to conform with the protocols of Poetry. The protocols lacked the user-friendliness of Prose.

Most individual Ancient Greek Homo Saps, their Thought processing focused on the reality of their Earthly environment, chose to adopt and develop the new prose application of their user-friendly Ancient Greek version of WritingITIS. Prose allowed individual Homo Saps to capture the Thought structures they sequenced in the free space and privacy of their individual Homo Sap memories without the obfuscations of the protocols of poetry and reproduce them in individual WritingITIS products.

PRINCIPAL HOMO SAP THOUGHT PROCESSORS AND HISTORIANS

The Homo Sap Thales was the first of individual Ancient Greek Homo Sap Thought processors who storytellers labelled as "philosophers." They were also Thought historians. Their developments of the Prose application of user-friendly Ancient Greek WritingITIS allowed them to process and re-sequence Mesopotamian and Ancient Egyptian Thought structures to eliminate all IS components describing gods or institutions as the originators of the Thought structures. The Thought structures that the Homo Sap Thales sequenced initialized the definition of the

COS and EOS as independent systems free of any godly or institutional controls.

The 15 Homo Sap generations of Ancient Greek philosophers after that of the Homo Sap Thales, captured the IS components in Thales' Thought structures, modifying them, refuting them, refining them, or expanding them with their own Thought structures. They followed his lead in processing and re-sequencing older Thought structures, stripping them of their godly and institutional control messages, to produce outcomes that described their Earthly realities and the power of individual Homo Sap Thought processing. Storytellers named these Ancient Greek Homo Sap Thought processors as "historians": Anaximander, Anaximenes, Heraclitus, Pythagoras, Empedocles, Xenophanes, Parmenides, Democritus, Melissus, Zeno, Anaxagoras, Leucippus, Protagoras, Prodicus, Hippeas, Georgias, Socrates, Plato, and Aristotle.

Storytellers labelled as "doctrines" the sets of Thought structures that the Ancient Greek philosophers sequenced and reproduced in WritingITIS products using the user-friendly Ancient Greek version of AlphabetITIS. Homo Sap Thought processors who operated in the old, established Ancient Egyptian and Mesopotamian Homo Sap civilizations had already captured most of the doctrines in their WritingITIS products, but obscured the essential messages of the doctrines in the institutional and sacerdotal context of their hieroglyphic and cuneiform ImageITIS products. The user-friendly Ancient Greek version of AlphabetITIS allowed the Ancient Greek philosophers to separate the doctrines from their institutional and sacerdotal context. They re-initialized the doctrines in simple Thought structures whose IS components defined them as the kernels of OS's controlling real Earthly and cosmic systems, and Homo Sap social and political systems.

Those who operated before the Homo Sap Socrates focused mostly on developing Thought structures whose IS components and outcomes focused on explaining their real physical environment of Earth and the Cosmos. One, the Homo Sap Pythagoras, sequenced Thought structures whose IS components disseminated messages that initialized the application of "numbers" as an ITIS product capable of capturing and describing the COS and EOS. The Pythagorean Thought structures initialized the protocols of number systems that Homo Saps labelled "mathematics," protocols for defining the kernel and layers of operating utilities of the COS and EOS.

They all demonstrated their skills as ITISans, the skills the outcomes of their free-Thought processing and their development of the user-friendly Ancient Greek versions of OralITIS and WritingITIS. Homo Saps labelled one group of philosophers as "sophists." They focused their ITIS products on the development of ITIS applications for the Ancient Greek versions of OralITIS and WritingITIS. Storytellers labelled one "Rhetoric," primarily an application of OralITIS. They expanded the prose application of WritingITIS to capture rhetoric in WritingITIS products.

The sophist philosophers also angered those Ancient Greek Homo Saps, the members of the governing democratic institution, who chose to develop Thought structures modifying the OS for their democracy. Their anger against the sophists demonstrated the insidious trap that individual Homo Saps designed for themselves when they joined institutions and confined their Thought processing to institutional Thoughtbases. The ITIS products of the sophist Homo Sap philosophers contained Thought structures that described procedures for analyzing existing Thought structures and re-sequencing them to produce new outcomes and IS components that criticized those of the new democratic institutions and OS.

Their freedoms from institutional or godly Thought-processing restrictions forced the Ancient Greek Homo Sap philosophers to invent new word components for their OralITIS and WritingITIS products, words that allowed them to capture the new concepts they described in the IS components of their new Thought structures. The Homo Sap Prodicus produced series of Thought structures focused on analyzing the development of their OralITIS products, the sounds that they produced with their hominid animal vocal technology to deliver the OralITIS products, the sequence of words for developing sentence structures that delivered consistent messages in both OralITIS and WritingITIS products, and the correct sequence of letters for spelling words. Storytellers labelled the Homo Sap Prodicus a "linguist."

The Homo Sap Protagoras produced Thought structures that initialized the study and science of new communications protocols of grammar, a critical ITIS development focused on ensuring that other Homo Saps captured the same messages from WritingITIS products.

The Homo Sap Protagoras identified different kinds of sentences. He developed procedures for developing words identifying male and female subjects of OralITIS and WritingITIS products

The Homo Sap Antiphon developed procedures for adapting the application of Rhetoric to the production of OralITIS products for other Homo Saps to deliver. Storytellers describe the Homo Sap Antiphon as the first "speech writer." He claimed to be able to produce OralITIS products whose IS components delivered messages focused on freeing any Homo Sap from sorrow.

The Homo Sap Anaxagoras sequenced Thought structures whose outcomes initialized the concept of Mind as a unique Homo Sap attribute. His Thought structures prompted other Homo Saps to describe Mind as a combination of individual Homo Sap memory, Thought-processing facility, and access node to the infinite Thought store of MEMORY.

Most of these Ancient Greek Homo Sap philosophers captured their Thought structures using the Prose application of Ancient Greek WritingITIS. The user-friendliness of Prose allowed individual Homo Saps to produce facsimiles of their vernacular OralITIS products in WritingITIS products. Using the vernacular feature of prose allowed individual Homo Saps to focus and sequence Thought into structures whose outcomes described their immediate reality, free of the unreal messages of the protocols of Poetry. They demonstrated an important feature of the Prose application of WritingITIS. The feature allowed individual Homo Saps to capture, sort, and isolate Thought structures that focused on single outcomes and only on the outcomes without obscuring their essential messages with conditional or optional outcomes.

Individual Homo Saps used a variety of protocols to capture and condition the messages in the Thought structures they captured from the ITIS products of other Homo Saps and then to disseminate their own versions of the messages. The Ancient Greek Homo Sap philosophers determined that the differences in these protocols caused the corruptions of the Thought sequences, IS components, and messages when they transmitted them from one Homo Sap to another.

To resolve the protocol problems, the Homo Saps Socrates, Plato, and Aristotle developed sets of Thought-processing exercises. They demonstrated a set of new protocols that all Homo Saps could use to clarify and disseminate exact facsimiles of their Thought structures, IS components, and outcomes in their WritingITIS products. Storytellers labeled these protocols "logic," "dialectic," "syllogism," "syntax," "categorization," and "classification." They initialized the new protocols as a set of IT specifications that they designed to function as ITIS standards for Homo Saps to adopt when sequencing the ITIS elements of their OralITIS and WritingITIS products. Their purpose was to ensure that other Homo Saps captured the correct IS components, messages, and outcomes of individual Homo Sap Thought processing.

They developed methodologies for defining the operating procedures for ITIS systems that delivered specific outcomes. Storytellers labelled the methodologies "induction," and "deduction." One hundred and twenty-five generations after that of the Homo Sap Aristotle, other Homo Saps relabelled the methodologies as "bottom-up" and "top-down" analysis technologies for the specifications of ITIS systems whose primary purpose was the storage and retrieval of Thought structures, their IS components and messages.

The Ancient Greek Homo Saps' Prose developments allowed the Homo Sap Plato, the ultimate Thought historian, to compile a Thoughtbase with Thought structures whose IS components disseminated messages that defined the conflict between individual Homo Saps and their Earthly institutions.

Plato expanded the messages that the Homo Sap Heraclitus first initialized in his Thought structures describing the antipathy that institutions expressed toward independent Thought processing and sequencing in individual Homo Sap memories. The Poetry application of the Ancient Greek WritingITIS products of the Homo Saps Homer and Hesiod contained Thought structures with IS components disseminating similar messages, but they identified the conflict as one outcome of the relationship between Ancient Greek Homo Sap Heroes and their patron gods, an unreal poetic relationship. The Homo Sap Plato's messages focused on Earthbound Homo Saps and Earthbound institutions that Homo Saps agreed to establish.

The Homo Sap Plato compiled a WritingITIS product with Thought structures whose IS components made real the conflict between the controls of institutional Thought processing and the freedoms of Thought processing in the free space of individual Homo Sap memories. His Thought structures made real the fears of the Ancient Egyptian god-king Narmer over the cosmic god Thoth's invention of the ITIS tool of WritingITIS.

Plato's Thought structures describe the Ancient Greek governing institution's attack on the Homo Sap Socrates for sequencing Thought structures with IS components disseminating controversial messages. When individual Homo Saps processed them in the privacy and free space of individual Homo Sap memories they prompted individual Homo Saps to produce outcomes that challenged the institutional control procedures of the Ancient Greek Homo Sap civilization.

The governing institution, though democratic, included a group of Ancient Greek Homo Saps that storytellers label "oligarchs." They described them as survivors of ancient institutions, Homo Saps who operated primarily in the Thought processing confines of old institutional Thoughtbases. The oligarchs convinced other members of the governing democratic institution to direct the Homo Sap Socrates to stop his Thought processing and disseminating activities.

The Homo Sap Socrates refused. They threatened him with HAP killing procedures. He chose to destroy his own HAP structure to disseminate the message declaring the priority of the Homo Sap Thought-processing imperative over institutional Thought-processing directives or the HAP survival imperative. The Homo Sap Socrates refused to compromise his individual Thought-processing freedom and submit to the Thought-processing controls of the new, though democratic, Ancient Greek governing institutions.

The Homo Sap Plato also compiled Prose WritingITIS products with Thought structures whose IS components disseminated messages that focused on the importance of privileged institutions in ensuring the orderly development of Homo Sap civilizations. The messages contradicted those he disseminated describing the priority of individual Homo Sap Thought processing over institutional Thought processing. The WritingITIS was primarily a summary history of the institutions other Homo Sap civilizations had devised to stabilize and control those civilizations.

The Homo Sap Plato initialized the outline of a new Thought structure with these prose WritingITIS products, which were all focused on the Earthly reality of Homo Saps free of gods. The IS component identified three distinct species co-existing in each individual Homo Sap: a hominid animal species, an individual Thought species, and an institutional Thought species. Each species operated to comply with the specific outcomes of three Thoughtbases that each individual Homo Sap stored in individual Homo Sap memories. Each Thoughtbase stored Thought structures focused exclusively on the survival and dominance of each species.

The Homo Sap Aristotle, developing still further the Prose application of Ancient Greek WritingITIS, produced WritingITIS products with Thought structures demonstrating the procedures and protocols for specifying the reality of the Earthly environment. He sequenced new words with the elements of the Ancient Greek AlphabetITIS to capture and describe the IS components of the new Thought structures specifying the Earthly reality that he and his Ancient Greek Homo Sap philosophers had identified.

The Ancient Greek Homo Saps disseminated their new Thought structures and their user-friendly ITIS developments to Homo Saps who operated in the Earthly region east and west of the Earthly region of Ancient Greece. The HAP survival imperative of the Ancient Greek king prompted the dissemination to the east. The HAP survival imperative of migrating Ancient Greek Homo Saps prompted the dissemination to the west.

Storytellers gave the name "Alexander" to the Ancient Greek king. The Homo Sap Aristotle personally transmitted the new Thought structures to Alexander. He initiated the dissemination to the east when he led his military institution in the execution procedures to subjugate the Homo Sap groups which operated in the Homo Sap civilizations of Ancient Egypt, Mesopotamia, and the northwest Earthly region of Ancient India. This dissemination made only minor changes to the institutionally controlled ITIS developments and Thought processing in the three Homo Sap civilizations.

After 125 Homo Sap generations of controlled Thought processing and ITIS developments in the Ancient Egyptian and Mesopotamian Homo Sap civilizations, the old institutions had made themselves essential to the continued viability of the civilizations. The outcomes of any major changes in their OS, Thought processing, and ITIS products would have produced chaos and destroyed the civilizations. In addition, the Homo Sap groups which operated in the eastern Homo Sap civilizations had developed compatible operating procedures whose outcomes allowed individual Homo Saps to function safely in the controlled safe havens' OS. Any changes to the established institutional order threatened their individual HAP safety as the institutions modified and enforced new OS to include procedures compatible with the new Ancient Greek Thought structures and user-friendly ITIS tools.

Most of the Earthly regions bordering the Mediterranean Sea to the west remained open to the Ancient Greeks' development of new safe havens. They needed new safe havens to accommodate the outcomes of their highly productive HAP sexual activities. The increasing numbers of individual Ancient Greek Homo Saps operating in the Ancient Greek safe havens began to overwhelm their available HAP survival resources. Many formed groups which chose to establish new safe havens outside the boundaries of the Ancient Greek Homo Sap civilization. Homo Saps labelled these new safe havens "colonies." The Ancient Greek Homo Sap colonists disseminated their new Thought structures and ITIS tools to the Homo Sap groups with which they interacted. They established stable safe havens in the Earthly region that storytellers labelled "Italy."

Their new Thought structures and ITIS tools had the greatest impact on the Homo Sap groups that storytellers labelled "Roman." The Roman Homo Saps modified the Ancient Greek user-friendly ITIS tools to produce their own versions and re-sequenced the new Thought structures of the Ancient Greek philosophers to develop their own Homo Sap civilization.

After the Homo Sap Aristotle died, Homo Saps of the following 125 Homo Sap generations reprocessed his and the Homo Sap Plato's Thought structures, IS components, and messages using adaptations of the user-friendly Ancient Greek versions of the ITIS tools of AlphabetITIS and WritingITIS. The conflict between individual Thought species and the institutional Thought species continued with the institutional Thought species periodically executing HAP killing procedures to assert their dominance over the individual Thought species. Storytellers describe the conflict as the Earthly manifestation of the conflict between to cosmic gods Ra and Thoth.

Under the direction of the cosmic god Thoth, Homo Saps continued to design ITIS developments that made the ITIS tools more user-friendly to individual Homo Saps. These developments always preceded Earthly developments whose outcomes gave more control over the operation of Homo Sap civilizations to the individual Thought species. Under Ra's direction, those Homo Saps in which the institutional Thought species dominated their actions, always executed HAP killing procedures to suppress Homo Saps' development of user-friendly ITIS tools.

Storytellers of other Homo Sap groups captured and reproduced the new Thoughtbases, often modifying and expanding the Ancient Greek Thought structures. The messages directed the Thought processing of 125 Homo Sap generations who formed themselves into a new Homo Sap civilization. The primary connection between the diverse Homo Sap groups who chose to join the civilization was the freedom to process Thought in individual Homo Sap memories. Storytellers labelled the civilization "Western."

AFTERWORD

Dear Jim:

Part Two of Our History of ITIS outlines the development and impact of ITIS tools, ITIS products, and their IT peripheral devices, on Homo Sap civilizations over the 125 Homo Sap generations after that of the Ancient Greek Homo Sap Aristotle. The Thought structures and their IS components will identify the strong connections between the freedoms that individual Homo Saps acquire and the development of user-friendly ITIS tools. Virtually all major changes—often violent—that individual Homo Saps or their institutions implemented in their OS followed the development of more and more user-friendly ITIS tools. Either the institutions initiated the changes to stop individual Homo Saps from acquiring skills in the applications of the ITIS tools or individual Homo Saps initiated the changes as they instinctively adopted user-friendly ITIS tools and began challenging moribund institutions over the control of Thought processing or their OS.

The Homo Sap groups of those generations formed three broad-based Homo Sap civilizations.

One civilization storytellers labelled "Middle Eastern" to incorporate the Homo Sap groups which operated in the Earthly regions of the Ancient Egyptian and Mesopotamian Homo Sap civilizations. The second storytellers labelled "Eastern" to incorporate the Homo Sap groups which operated in the Earthly regions of the Ancient Indian and Chinese Homo Sap civilizations. The third civilization storytellers labelled "Western" to incorporate the Homo Sap groups which operated in the Earthly regions west of Ancient Greece.

Storytellers, their numbers increasing rapidly as more and more individual Homo Saps acquired ITIS skills, captured in WritingITIS products Thought structures with IS components containing powerful and persistent messages. One set of messages identified ITIS developments in the Western Homo Sap civilizations as the primary events whose outcomes gave individual Homo Saps more and more influence on the designs of the OS of their safe havens and on their relationships with institutions that operated in the safe havens.

The messages, many from storytellers who operated only as institutional Thought species, invariably followed Homo Saps' initialization of new ITIS tools and IT peripherals with features that made them more user-friendly for individual Homo Saps to acquire skills in their applications without interference from their institutions. Individual Homo Saps acquired more freedoms to challenge the Thought-processing controls of the institutions with every increase in the user-friendliness of the ITIS tools. The institutional paradox was that individual Homo Saps quickly adopted and adapted to enhance their individual Thought-processing freedoms the ITIS tools whose development institutions sponsored to reinforce their Thought-processing control.

The messages described the speed with which individual Homo Saps adopted and adapted ITIS tools as "phenomenal," equivalent to the speed with which they adopted and implemented new HAP survival procedures. The Thought structures describe the consummate skills individual Homo Saps acquire in ITIS applications as "instinctive." The outcomes demonstrated the power of the Homo Saps Thought-processing imperative to identify the most user-friendly ITIS tools and to direct individual Homo Saps to immediately acquire ITIS skills primarily to access the infinite Thought store of MEMORY, but also to overcome institutional impediments to free-Thought-processing in individual Homo Sap memories.

Homo Saps who chose to operate in the Middle Eastern and two Eastern Homo Sap civilizations initialized versions of AlphabetITIS and other ITIS tools and peripherals but they failed to develop them. The dominant institutions of the civilizations, the Brahmins in the Indian Homo Sap civilization, and the Mandarins in the Chinese Homo Sap civilization, maintained absolute control of ITIS developments, sanctioning only those developments that enhanced institutional hegemony.

Many of the polyglot Homo Sap groups which operated in the Middle Eastern civilization implemented versions of the ITIS tools for interactions between individual Homo Saps who belonged to the same groups and developed versions of AlphabetITIS but they had little impact on the OS and procedures of the collective of Homo Sap groups that formed the civilization.

Homo Saps who operated in the Eastern Homo Sap civilization, primarily that of the Chinese Homo Sap groups, developed many IT peripherals before the Western Homo Saps developed similar ones—but they failed to develop them. The Chinese institutions sanctioned their development only for godly, aristocratic, or artistic applications.

Homo Saps of the first 15 generations after that of the Homo Sap Aristotle transmitted their philosophical Thought structures and demonstrated the applications of their user-friendly ITIS tools of AlphabetITIS and WritingITIS to Homo Sap groups that storytellers labelled "Roman." The Roman Homo Saps adapted the Ancient Greek AlphabetITIS, adding new simple ImageITIS elements to represent the sounds they produced with their HAP vocal technology to implement their version of OralITIS. Storytellers labelled the Roman Homo Saps' version of OralITIS "Latin."

The Roman Homo Saps captured the Ancient Greek Thought structures and began to focus their ITIS developments on designing practical applications of the messages and procedures in the Ancient Greek Thought structures. They implemented their ITIS applications to form the OS of the Western Homo Sap civilization. As with earlier Homo Sap civilizations, except that of the Ancient Greeks, three institutions: aristocratic, military, and priestly, implemented different sub-systems of the OS, but all under the control of the aristocratic institution which operated the kernel procedures and nominated the leaders of their military and priestly institutions. The priestly Homo Saps implemented applications of the Ancient Roman version of the ITIS tool of CalendarITIS and arbitrarily modified sequences of political, priestly, and social events in the Ancient Roman safe havens, generally at the whim of the dominant aristocratic Homo Sap faction directing the actions of the aristocratic institution.

Storytellers have focused many WritingITIS products on Thought structures with IS components disseminating messages describing the character of the leader who assumed the direction and then initialized new Thought structures to begin the redirection of the Ancient Roman Homo Sap civilization. They gave him the name "Julius Caesar."

The messages identify the Homo Sap Julius Caesar as the complete ITISan. He demonstrated consummate skills in the applications of the

ITIS tools of OralITIS, ImageITIS, and WritingITIS. He developed and imposed a new version of CalendarITIS on the Roman Homo Saps, an adaptation of the Ancient Egyptian civic version of CalendarITIS. The new CalendarITIS specified a predictable fixed order to events in the Roman Homo Sap civilization and eliminated the freedom of the aristocratic and priestly institutions to arbitrarily control the sequencing of events in the Ancient Roman safe havens.

He applied his ITIS skills to initiate and complete major changes in the OS of the Ancient Roman safe havens. He curtailed the excesses of the Homo Sap members of the aristocratic institution. Though he was the leader of all four Ancient Roman institutions—aristocratic, priestly, military and governing senate—he refused the demands of the institution members to restore their traditional aristocratic freedoms to operate without restrictions on their control of the Roman OS.

The messages describe his individual freedom to operate at will as an individual Homo Sap Thought species, an institutional Thought species, and a hominid animal species, choosing to operate and process Thought structures of each Thoughtbase to achieve specific outcomes. Many of the outcomes disrupted the traditional controls of the aristocratic institution and prompted an aristocratic Homo Sap faction, all of whose members operated only as institutional Thought species, to assassinate him.

Storytellers produced many WritingITIS products containing Thought structures describing the ITIS developments of another Ancient Roman Homo Sap with ITIS skills who operated during the same Homo Sap generation as that of Julius Caesar. They gave him the name "Cicero."

The Homo Sap Cicero was a leading member of the aristocratic institution who demonstrated consummate skills in the Prose applications of WritingITIS. Storytellers describe him as having great artistic skills in the sequencing and delivery of the Thought structures of his OralITIS and WritingITIS products. Some storytelling Thought structures attribute the Homo Sap Cicero as the demonstrator of the Ancient Greek Homo Sap Simonides' methodology for storing Thought structures in individual Homo Sap memories and retrieving them instantly at will.

At the end of 15 Homo Sap generations after that of the Homo Sap Julius Caesar, the Ancient Roman aristocratic institution lost control of the Thought processing in the Western Homo Sap civilization. A new priestly institution began to establish control of Thought processing and ITIS developments in the Western Homo Sap civilization.

Three Homo Saps initialized the Thought structures around which they began to design a Thoughtbase and the kernel procedures of a new OS for the Western Homo Sap civilization. Storytellers gave them the names "Jerome," "Ambrose," and "Augustine." They labelled the IS components of the controlling Thought structures of the new institution and the messages they disseminated as "Christianity." Storytellers labelled as "Christians" all the Homo Saps who chose to process Thought structures only from the Christianity Thoughtbase.

One Homo Sap member of the priestly institution began the process of capturing all OralITIS and WritingITIS products containing Thought structures, IS components, and messages that impacted those in the Christianity Thoughtbase. Homo Saps gave him the name "Benedict."

The Homo Sap Benedict began gathering other Homo Saps who chose to operate in accordance with the procedures and messages of the Christianity Thoughtbase to form a subsidiary institution of the new priestly institution. Benedict directed the priestly Homo Saps to construct safe havens in which to produce food resources for the survival and sustenance of their HAP structures. As with many Thought structures that Ancient Greek and Ancient Roman Homo Saps processed, the initializing Thought structures for such safe havens were those that Ancient Egyptian Homo Saps who chose to operate exclusively in the Christianity Thoughtbase sequenced. Storytellers labelled this Ancient Egyptian Homo Sap group "Copts."

Storytellers labelled the priestly safe havens "monasteries" and the priestly Homo Saps who chose to operate in the safe havens "monks." They labelled these Homo Sap monks "Benedictines." The priestly Homo Sap Benedictine established for the Benedictine monks sets of rules of conduct and procedures to follow while they operated in the monasteries.

Benedictine directed his monks to acquire WritingITIS skills with which to capture in new WritingITIS products the Thought structures of the WritingITIS and ImageITIS products that other Homo Saps had produced before and during the development of the Christianity Thoughtbase. Some of the Homo Sap monks began to modify the Thought structures they copied to make the IS components and messages conform to those in the Christianity Thoughtbase. Other Homo Saps monks simply copied verbatim any WritingITIS product that their priestly leaders directed them to copy. They stored all WritingITIS products, originals and copies, in the ITIS storage peripherals of libraries. The Homo Sap monks restricted access to the ITIS products they deposited in the monastery libraries to Homo Sap monks who operated in the monasteries or to those priestly Homo Saps who operated as Christian Homo Saps outside the monasteries.

Itinerant priestly Homo Saps travelled around the Earthly region of the Western Homo Sap civilization. They dedicated their Earthly activities to disseminating OralITIS products they designed to disseminate the IS components, and messages of the Christianity Thoughtbase to convince other Homo Sap groups—storytellers labelled them "Pagan"—to begin processing Thought structures only from the Christianity Thoughtbase. The itinerant priestly Homo Saps had to develop and demonstrate their ITIS skills in making the Christianity messages clear to the pagan Homo Saps.

Some individual Homo Saps who the itinerant priestly Homo Saps induced to capture the Thought structures of the Christianity Thoughtbase began to acquire WritingITIS skills under the direction of the itinerant priestly Homo Saps; many of these individual Homo Saps belonged to new aristocratic institutions that the Homo Sap leaders of the Homo Sap groups which operated in the Western Homo Sap civilization began to form to replace the Roman Homo Saps aristocratic institution.

Five Homo Sap generations after that of the priestly Homo Sap Benedict, the Christian priestly and Western aristocratic institutions succeeded in imposing their Thought processing and HAP controls on the Homo Saps who operated in the Earthly region of the Western Homo Sap civilization. The controls allowed them to establish their domination of the Christian operating procedures, IS components, and messages over all those who the Western Homo Sap groups had adopted in the past.

At the end of the five Homo Sap generations, an individual Homo Sap who operated in the Earthly region of the Middle Eastern Homo Sap civilization sequenced a set of Thought structures specifying procedures, IS components, and messages that Middle Eastern Homo

Saps compiled as the kernel of an OS for a new priestly institution. Storytellers gave the name "Mohammed" to the individual Middle Eastern Homo Sap and labelled him a "Prophet."

The Middle Eastern Homo Saps who followed the Prophet Mohammed collected the Thought structures he sequenced in a Thoughtbase that they reproduced and disseminated in WritingITIS products. Storytellers labelled the outcome of the Thought structures "Islam." The leaders of the first few generations of Islam's adherents belonged to a Homo Saps groups that storytellers labelled "Arabs." The Thought structures they disseminated contained IS components with fundamental messages similar to those of the Christianity Thoughtbase. When the Arabs lost control of the Islamic messages, other Homo Saps of different groups who claimed to follow the Prophet Mohammed began to change the messages, making them ominous and frightening.

Five Homo Sap generations after that of the Prophet Mohammed, the Middle Eastern Homo Sap groups had imposed their Islamic OS on all Homo Sap groups of the Middle Eastern Homo Sap civilization. The Islamic Homo Saps expanded their civilization to the Earthly region that storytellers label "Spain" on the western border of the Western Homo Sap civilization.

The military institution of the western Homo Sap groups that storytellers labelled "Merovingian" stopped the Islamic Homo Saps from further expansion into the Western Homo Sap civilization. Storytellers gave the name "Charles Martel" to the leader of the Merovingian Homo Sap groups who led his military institution in stopping the advance of the Islamic Homo Saps into the Western Homo Sap civilization.

The Christian priestly institution, as adamant as the Islamic priestly institution in declaring the primacy of one religious Thoughtbase over the other, vigorously opposed the dissemination of Islamic Thought structures and OS into the Western Homo Sap civilization. For the following 25 Homo Sap generations the Christian priestly institutions threatened individual western Homo Saps with godly retribution and damage to their HAP structures should they process or reproduce in ITIS products disseminating Islamic Thought structures, IS components, and messages.

Two kings of the Western Homo Sap civilization independently captured ITISanian Thought structures that the priestly Homo Saps disseminated with their institution's controlling Thought structures. Storytellers gave the names "Charles and Alfred" to the two kings and added IS components to make the names "Charlemagne" and "Alfred the Great." The messages in the IS components declared both as having superior Homo Sap leadership skills.

The ITISanian Thought structures they processed disseminated messages that described the power of the ITIS tool of WritingITIS both to control Thought processing in individual Homo Sap memories and to free individual Thought processing from the moribund routines of institutional Thoughtbases. Both Charlemagne and Alfred the Great sequenced Thought structures with IS components and messages directing individual Homo Saps in their Homo Sap groups to acquire ITIS skills.

Charlemagne and Alfred the Great sequenced and disseminated Thought structures with IS components and messages directing the priestly Homo Saps who operated in their safe havens to design and produce WritingITIS products with Thought structures that converted the Latin version of OralITIS into the OralITIS versions that their Homo Sap groups used to communicate. Storytellers labelled the new WritingITIS products "vernacular versions" of the Latin WritingITIS products that the priestly Homo Saps produced.

The ITIS directives and Homo Sap leadership of Charlemagne and Alfred the Great initiated a rapid increase in the numbers of individual Homo Saps who acquired WritingITIS skills. These individual Homo Saps began to function as storytellers, capturing in WritingITIS products Thought structures they sequenced with the ITIS elements of different vernacular versions of OralITIS.

Storytellers describe the outcome of these ITIS events, the outcome manifesting 15 Homo Sap generations after that of Alfred the Great, as an intense surge in individual Homo Sap Thought processing of Thought structures they sequenced in individual Homo Sap memories in and out of the priestly and aristocratic Thoughtbases to produce ITIS products capturing Thought structures with IS components and messages different from those that the priestly and aristocratic institutions produced.

Storytellers describe two new ITIS events that Homo Saps initiated after the surge in individual Thought processing. The priestly Homo Saps initiated one event when they initialized sets of Thought structures whose procedures they implemented as OS of subsidiary institutions. They dedicated these institutions to the control of the dissemination of new Thought structures with IS components and messages that threatened the stability of the Western Homo Sap civilization and of the priestly and aristocratic institutions. Storytellers labelled the subsidiary institutions "universities."

The procedures of the university OS, primarily sets of rigid protocols with fixed outcomes, directed the Homo Sap members of the universities, most of whom belonged to the Christian institutions and who storytellers labelled "professors." The directive required the Homo Sap professors to disseminate to other Homo Saps who storytellers labelled "students," only Thought structures with IS components, and messages that conformed to those that the priestly institution authorized. The directive also required the Homo Sap professors to focus their ITIS skills on discouraging or suppressing the development and dissemination of new Thought structures with IS components and messages that failed to meet the institutional standards. Storytellers describe the development of the first university institutions as the outcome of the priestly institution's tacit acknowledgement of institutional impotence in controlling independent Thought processing in individual Homo Sap memories or in controlling individual Homo Saps' independent acquisition of ITIS skills.

Two individual Homo Saps initiated the next ITIS events. The outcomes revitalized Homo Saps' focus on individual Homo Sap memories and MEMORY as ITIS peripherals, one for storing and retrieving Thought structures and the other as the store of the infinity of all Thought from which individual Homo Saps could retrieve Thought structures and sequence new Thought into new structures defining procedures, IS components, and messages that produced any outcome possible in the physical environment of Cosmos. Storytellers gave the names "Thomas Aquinas" to one and "Ramon Lull" to the other.

The Homo Sap Thomas Aquinas, a member of the sub-institution of the Christian institution that storytellers label "Dominican," adapted the individual Homo Sap memory Thought storage and retrieval methodology that the Ancient Greek Homo Sap Simonides described in his WritingITIS product. Thomas Aquinas' adaptation specified Christian ImageITIS products as the key Thought nodes for randomly accessing for instant retrieval the networks of Thought structures that individual Homo Saps stored in their individual Homo Sap memories. The Christian ImageITIS products helped keep individual Homo Saps operating in the Christianity Thoughtbase, ensuring that any modifications to messages conformed to those of the Christianity Thoughtbase.

The Homo Sap Ramon Lull, a member of the sub-institution of the Christian institution that storytellers label "Franciscan," designed ITIS peripherals that allowed individual Homo Saps to access MEMORY, the store of the infinity of all Thought and Thought structures. He identified Thoughtbases that defined the components and OS of the Cosmos as well as the godly and Earthly connections between Cosmos, gods, Homo Saps, and all living things. He designed his storage and retrieval system to conform to the requirements of three priestly institutions, Christian, Muslim, and Hebrew, each of which developed a Thoughtbase with Thought structures that the three sets of priestly Homo Saps retrieved from similar WritingITIS products, but each modified their institutional OS and Thoughtbases to produce different IS components, messages, and outcomes.

Ramon Lull initialized an IT that allowed individual Homo Saps to implement his retrieval methodology. The IT combined key Thought to retrieve Thought structures from MEMORY whose IS components, messages, and outcomes identified interconnections between any set of the infinity of Thoughtbases that MEMORY contained.

Homo Saps of the Western Homo Sap civilization began expanding the Thought structures of the priestly Homo Saps Thomas Aquinas and Ramon Lull 15 Homo Sap generations after their Homo Sap generation. Each set of Thought structures contained IT and IS components. One IT, that of the priestly Homo Sap Thomas Aquinas, specified a serial storage system, the retrieval methodology always producing an outcome compatible with the IS components and messages of the Christianity Thoughtbase. The second IT, that of the priestly Homo Sap Ramon Lull, specified a random access storage and retrieval system, the storage design and retrieval methodology producing outcomes compatible with the messages in three priestly Thoughtbases, those of Christianity, Islam, and Judaism.

Homo Saps' expansion of the system Thought structures included sequences that associated memory systems with the Earthly manifestation of the cosmic god Thoth. Storytellers gave the name

"Hermes Trismegistus" to Thoth's mysterious but Earthly counterpart. Storytellers labelled the IS components and messages in these expanded Thought structures "Hermetic" and the outcome of the Thoughtbase the "Hermetic Tradition." The Hermetic Thoughtbase stored Thought structures whose procedures, IS components, and messages identified many more sources of Thought structures explaining the COS than the single godly source that the Christian priestly institution prescribed.

Individual Western Homo Saps processed and disseminated the Hermetic Thought structures during an Earthly period spanning 15 Homo Sap generations that storytellers label "The Renaissance." Many individual Homo Saps focused their Thought processing on the Hermetic Thought structures and procedures that identified memory systems as critical ITIS devices. One Homo Sap to whom storytellers gave the name "Giullo Camillo," constructed a memory theater as an IT whose design allowed individual Homo Saps to retrieve the answer to any question. Another individual Homo Sap to whom storytellers gave the name "Giordano Bruno," a member of the Dominican sub-institution of the Christian priestly institution, designed a different memory system through which individual Homo Saps could also retrieve the answer to any question. The Christian priestly institution executed Giordano Bruno with HAP killing procedures, burning him alive, for developing and disseminating his Hermetic Thought structures.

An individual Homo Sap initiated a major ITIS event that radically changed Thought processing in the Western Homo Sap civilization. Storytellers gave the individual Homo Sap the name "Gutenberg." He installed the first printing press as an ITIS peripheral capable of producing and reproducing exact copies of one WritingITIS product. This production allowed many individual Homo Saps of the same generation to capture and process the same Thought structures in the privacy of their individual Homo Sap memories, individually modifying IS components and messages and re-sequencing Thought structures to produce outcomes different from those of their controlling priestly and aristocratic institutions.

The institutions attempted to end this great expansion in free-Thought processing with threats of damage to or destruction of the HAP structures of individual Homo Saps who produced or disseminated Thought structures with IS components and messages different from those of the institutions. Storytellers labelled the HAP killing procedures of the Christian priestly institution "the Inquisition." The ultimate outcome was a demonstration of the power of individual Homo Saps' Thought-processing imperative directing them to adopt user-friendly ITIS tools and develop their capabilities for free-Thought-processing. Within two Homo Sap generations after that of the Homo Sap Gutenberg—unprecedented speed for Homo Saps—other Homo Saps installed printing presses in virtually every safe haven in the Western Homo Sap civilization, openly defying their controlling institutions.

Storytellers gave the names "Descartes, Leibniz, and Newton" to the three individual Homo Saps in the last Homo Saps generation of the Renaissance whose individual Thought processing initiated a whole series of ITIS events. They sequenced and disseminated Thought structures whose outcomes, procedures, IS components, and messages began the massive expansion of Thought processing in individual Homo Saps of the next 15 Homo Sap generations of the Western Homo Sap civilization.

Each had captured and processed the Thought structures of the Hermetic Thoughtbase before they sequenced and disseminated Thought structures, IS components, and messages that began the redefinition of the COS and EOS. The priestly institutions and those storytellers who operated only as institutional Thought species labelled the Hermetic Tradition "occultism," "magic," and eventually "heretic." They forced those individual Homo Saps who exercised their freedom to process Hermetic Thought structures to confine the outcomes in the silence of their individual Homo Sap memories and disseminate them in secret conclaves in fear of the institutional threats of damage to their HAP structures.

For the last 15 Homo Sap generations of the Western Homo Sap civilization, individual Homo Saps redesigned the ITIS tools to make them more and more user-friendly. The outcomes prompted individual Homo Saps to continually challenge absolute institutional controls on Thought-processing. The user-friendly ITIS tools allowed many individual Homo Saps to focus their Thought processing on analyzing and debating the requirements and their priorities of the three species— individual Thought species, institutional Thought species, and hominid animal species—simultaneously operating in individual Homo Saps to develop and impact the development of Homo Sap civilizations.

The analyses and debate intensified with Homo Saps' development of the as-yet ultimate ITIS peripheral and Earth-changer. Storytellers labelled this ITIS peripheral "the Personal Computer." Individual Homo Saps acquired consummate skills in the applications of the Personal Computer in less than a Homo Sap generation, undermining the absolute powers of some institutions and causing many to collapse.

Homo Saps have yet to identify the ultimate outcome. The followers of the Hermetic Tradition claim that Hermes Trismegistus— the Earthly manifestation of the cosmic god Thoth—has identified but keeps secret the outcome until Homo Saps have developed the "wisdom" to implement the outcome without destroying themselves.

ABOUT THE AUTHOR

John Barber entered the ancient information technology and information systems business as a merchant marine radio officer. He travelled the world, crossed all five geographical circles, and then decided to obtain a couple of university degrees: a bachelor's in Applied Science and a master's in Communications. Since then, mostly as a freelance, he has immersed himself in the modern information technology and information systems business, mostly in Systems Analysis and information storage and retrieval systems. He has taught distance learning and on-line courses in Information Technology for B.C. Open University and presently tutors an on-line course in Information Technology for Thompson Rivers University, Open Learning Division. He is a grandfather and libertarian.